Environmental Epidemiology: Epidemiological Investigation of Community Environmental Health Problems

Editor

John R. Goldsmith, M.D., M.P.H.
Professor
Epidemiology and Health Services
Evaluation Unit
Faculty of Health Sciences
Ben Gurion University of the Negev
Beer Sheva, Israel

CRC Press
Taylor & Francis Group
Boca Raton London New York

CRC Press is an imprint of the
Taylor & Francis Group, an **informa** business

CRC Press
Taylor & Francis Group
6000 Broken Sound Parkway NW, Suite 300
Boca Raton, FL 33487-2742

Reissued 2019 by CRC Press

A Library of Congress record exists under LC control number:

Publisher's Note
The publisher has gone to great lengths to ensure the quality of this reprint but points out that some imperfections in the original copies may be apparent.

Disclaimer
The publisher has made every effort to trace copyright holders and welcomes correspondence from those they have been unable to contact.

ISBN 13: 978-0-367-20685-7 (hbk)
ISBN 13: 978-0-367-20687-1 (pbk)
ISBN 13: 978-0-429-26288-3 (ebk)

Visit the Taylor & Francis Web site at http://www.taylorandfrancis.com and the
CRC Press Web site at http://www.crcpress.com

PREFACE

To our readers — our students:

Our goal of this book is to tell the story of how various persons and groups have successfully dealt with a type of problem which may threaten the lives and health of every group of humans — every community. The problem is that of a polluted environment.

This is *not a textbook of the usual kind* in the sense that it expounds a single body of knowledge and the principles on which it is based.

It is a textbook of a new and different kind in the sense that it provides systematic information for dealing with an increasingly important set of problems based on the experiences of many communities and the scientists involved in collecting and analyzing the data relevant to these problems.

The scholarly student who studies this book will gain more from it if she or he already knows, studies, or has available a background in the three relevant disciplines of epidemiology (including biostatistics and survey methods); environmental physics, chemistry, and biology (environmental science); and toxicology, whose applications are involved in community studies of environmental health. But the plan and style of the book are intended primarily for the student with a general scientific background or with a background in public administration at the undergraduate level.

The ardent student who is outraged by the apparent impotence of "advanced" societies to protect its children from poisons and unrestrained automobile traffic, or its asthmatics from breathtaking pollution will find here encouragement from the record of the limited successes achieved by scientists and communities when they work smoothly together.

The student activist who wants to participate in protecting the health of his community will find here tools suitable for a diversity of problems of which pollution is but one.

The ordinary student to whom this book is dedicated will find an introduction to certain branches of applied science and to the relationships between science and government.

This is not a "how to do it" book; it is a "what can be done" book, based on grappling with real problems; in most cases, the story is told by the scientists who participated in shedding light on them. There are many good textbooks in the three disciplines of epidemiology, environmental science, and toxicology, and we shall mention some later. Yet, these disciplines are difficult ones to learn in an efficient way by any method short of supervised work on real problems. Opportunities for such experience in dealing with community problems in environmental health are scarce in relation to the need for such teaching. We hope that the examples given in the second section of the book will be clear enough and challenging enough to simulate personal experience, and that the first and third sections along with exercises suggested by them and your teachers will begin to recreate the sense of personal challenge that the authors have felt and earnestly want to transmit. As with sex and mountain climbing, nothing quite substitutes for personal experiences, but people write books about those subjects too.

Books about the coming environmental disasters get to be best sellers, and books about how to manipulate your Karma seem to be on many bedside tables. A book on how facts can be used to influence communities so as to avert disaster and related threats seems overdue, but you, our students must on the basis of your own insights and actions write such a book; we are only able to start you off with the accounts of how other communities and investigators have tackled their problems and what came of the effort.

Although for purposes of teaching and analysis we maintain the fiction that each community and its problems stands alone, on the basis of modern concepts of ecology, we know that this is a false assumption. Each community which uses knowledge to make its citizens' lives safer or more healthful adds to the dignity of every man and increases the appreciation of the human potential and the goodness of nature.

John R. Goldsmith

ACKNOWLEDGMENTS

Most of the writing of this book took place during a half-year sabbatical leave spent at the University of Michigan School of Public Health in Ann Arbor. Professor Victor Hawthorne, Professor Ian Higgins, and all of the staff members of the Department of Epidemiology were most generous with their time. Access to the excellent library of the school made it relatively easy to check on important facts and citations. Special thanks are due David Hunsche and Mary Weed of the Department of Audio-Visual Services who never seem to run out of time or patience.

Professor Ido DeGroot of the Planning and Epidemiology Units, University of Cincinnati, Professor Tee Guidotti, University of Alberta School of Medicine, Mr. William Rokaw, U.S. Centers for Disease Control, Dr. Jill Joseph, University of Michigan School of Public Health, Marie Haring Sweeney of the National Institute of Occupational Safety and Health, Cincinnati have read, commented, and suggested improvements for which I am grateful.

Much of the experience on which the book is based was acquired during years spent at the California State Department of Public Health, and to my colleagues there and those who identified with the Department in its efforts to improve health in California I am greatly indebted.

Chapter 17 was included as a direct result of a discussion of the book with Dr. William Haddon, Jr., President, Insurance Institute for Highway Safety, whose death in March of 1985 deprives public health of a great leader.

THE EDITOR

John R. Goldsmith, M.D., M.P.H. is Professor of Epidemiology at the Faculty of Health Sciences, Ben Gurion University of the Negev, Beer Sheva Israel.

Professor Goldsmith graduated from Reed College, Portland, Oregon in Physics, and obtained his training in Medicine and Public Health at Harvard in 1942 to 1945 and 1955 to 1957, respectively. He served an internship at the University of Chicago Clinics and a Residency in Medicine at the University of Washington, Seattle, where he began his research on diseases of the lung. He was a family physician for five years in Salem, Oregon. From 1957 to 1979 he was in charge of research on the health effects of air pollution at the California State Health Department, and was responsible for the scientific bases of the California Air Quality Standards. He was seconded by the U.S. Public Health Service to the Headquarters of the World Health Organization in Geneva, Switzerland from 1964 to 1966, where he worked as an environmental epidemiologist. From 1973 to 1975 he was assigned to the Office of the Associate Director for Field Studies and Statistics of the U.S. National Cancer Institute in Bethesda, Maryland. There his work was on epidemiological study of environmental factors in cancer. He moved to Beer Sheva, Israel in 1978, where in addition to teaching and consulting, his research is divided between environmental health and evaluation of changes in health care.

Professor Goldsmith is a member of the American Association for the Advancement of Science, the American Thoracic Society, the International Epidemiological Association, the Society for Epidemiological Research, the Permanent Commission for Occupational Health, and a Fellow of the American Public Health Association. He was a member of the Air Conservation Commission of the American Association for the Advancement of Science, served on many committees of the American Thoracic Society and the American Lung Association, including Chairmanship of the Committee on the Health Effects of Air Pollution; he serves on the Scientific Committee on Epidemiology in Occupational Health for the Permanent Commission for Occupational Health; he has served on numerous panels and committees of the National Academy of Sciences-National Research Council of the U.S. and is a member of the Scientific Committee on Air Quality Standards for the Environmental Protection Service of Israel. He has served as a reviewer of research proposals for the National Institutes of Health, the Environmental Protection Administration of the U.S., and the American Lung Association. He is a former member of the U.S. National Committee on Vital and Health Statistics, for whom he headed the Panel on Statistics needed to determine the Effects of Environmental Exposures on Health. He is on the editorial boards of numerous scientific journals and has published scores of articles, reports and analyses.

Professor Goldsmith's recent publications are on epidemiological monitoring for environmental health protection, on methods for multivariate analyses, on strategies for pooling data, and on effects of environmental exposures on reproductive outcomes.

CONTRIBUTORS

Margaret Deane, M.P.H.
Epidemiological Studies Section
California State Department of Health
  Services
Berkeley, California

Roger Detels, M.D.
Dean and Professor of Epidemiology
School of Public Health
University of California at Los Angeles
Los Angeles, California

Badri Fattal, Ph.D.
Environmental Health Laboratory
Hebrew University — Hadassah Medical
  School
Jerusalem, Israel

Ron G. Frezieres, M.S.P.H.
Project Director — Research Unit
Los Angeles Regional Family Planning
  Council
Los Angeles, California

John R. Goldsmith, M.D., M.P.H.
Professor
Epidemiology and Health Evaluation
  Unit, Faculty of Health Sciences
Ben-Gurion University of the Negev
Beer-Sheva, Israel

Inge F. Goldstein, Dr.PH.
Clinical Associate Professor, Director
Environmental Epidemiology Research
  Unit
Columbia University School of Public
  Health
New York, New York

Martin Goldstein, Ph.D.
Professor of Chemistry
Yeshiva University
New York, New York

Diana Hartel, M.P.H.
Environmental Epidemiology Research
  Unit
Columbia University School of Public
  Health
New York, New York

Roger M. Katz, M.D.
Clinical Professor
Department of Pediatrics
University of California at Los Angeles
  School of Medicine
Los Angeles, California

Herbert L. Needleman, M.D.
Department of Psychiatry
Children's Hospital
Pittsburgh, Pennsylvania

Gad Potashnik, M.D.
Senior Lecturer
Department of Obstetrics and Gynecology
Soroka Medical Center
Beer-Sheba, Israel

Allen B. Rice
Michigan Council on Alcohol Problems
Lansing, Michigan

Stanley N. Rokaw, M.D.
American Lung Association of Los
  Angeles County
Los Angeles, California

Hillel I. Shuval, M.P.H.
Professor and Director
Environmental Health Laboratory
School of Public Health and Community
  Medicine
Hebrew University — Hadassah Medical
  School
Jerusalem, Israel

Toshio Toyama, M.D.
Professor Emeritus
Department of the Preventive Medicine
 and Public Health
Keio University School of Medicine
Tokyo, Japan

Alexander C. Wagenaar, Ph.D.
University of Michigan Transportation
 Research Institute
Ann Arbor, Michigan

Perez Yekutiel, M.D.
Visiting Professor
Environmental Health Laboratory
Hebrew University — Hadassah Medical
 School
Jerusalem, Israel

TABLE OF CONTENTS

Section III: Environmental Epidemiological Studies as a Basis for Health Protection

Section I: Background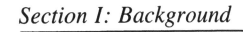

Chapter 1

COMMUNITIES: THEIR STUDY AND THEIR ENVIRONMENTAL PROBLEMS

John R. Goldsmith

TABLE OF CONTENTS

I. HOW COMMUNITIES DIFFER FROM INDIVIDUALS

Although a community is a difficult thing to define, it is easy to define what is *not* a community. No single person, no nuclear family, no bureaucracy, no government, no corporation, no place, and no process can be considered as a community. Each person *belongs to,* each family *is important to,* each effective bureaucracy *is beholden to,* each democratic government *must be responsive to,* and many corporations *affect the lives of* communities.

A community is then a group of living organisms with a shared location, shared environment, and shared fate.

A human community may or may not be organized, have selfawareness, selfgovernment, and selfconfidence. Communities are often organized, gain selfconfidence and selfesteem *because of their capabilities to respond effectively to threats to their members.* Environmental threats are an important class of such threats.

Most of the time most of us are in good health, but from time to time this is not the case, and we may say we are sick. We may have a cold, a stomach ache, a fever, arthritis, a tumor, a broken arm, a burn, and of course many more conditions. When we are sick, several things follow. We often are excused from work or school; if we are not literally absent, then we may excuse a poor performance because we are sick. We may ask a doctor for treatment, we may wait and hope to recover with no treatment, or we may treat ourselves. When sick, what we most want to know is how serious this illness is, then what is the cause and what to do to recover. We, the sick individuals, usually take the responsibility of initiating action, but we expect that the community will be supportive, and that our immediate family members will be most supportive of all the members of the community.

Communities also vary in their state of well being. Community diagnosis is one of the central concepts in current developments in academic departments of community medicine. The experiences of communities in recognizing and responding to environmental threats amply illustrate the validity of this concept. Yet for the most part this experience is not taken advantage of. Other types of community "dis-ease" are given priority.

Unemployment at a local factory or mine may abruptly reduce the income and erode the selfesteem of many members of the community. A crime wave, delinquent gangs, or inadequate housing, educational, recreational, or child care facilities may make living in the community unattractive. Floods, blizzards, earthquakes, or drought may make some communities "ill" or "injure" them so that they take months or years to recover. In general, people want to move into a thriving community and want to leave a "sick" one. The same quality of housing is likely to cost much more in a healthy or thriving community than it is in a sick one. The diagnosis, treatment, and outlook for recovery of community sickness is not as clear as that for individuals — and all too often the diagnosis and management of sickness in individuals is far from satisfactory!

We have mentioned the economic, social, and natural disasters which may impair the health of a community as well as the lack of needed resources. One of the increasingly common types of community illness is environmental. Inadequate housing, a contaminated water supply, noise from freeways or airports, or community air pollution have increasingly impaired the health of communities, making them less desirable places in which to live and work, depressing prices of housing, and discouraging immigration. Sometimes the community malady is manifested by sickness or death of its members, but more frequently, the environmental health impact on its residents is subtle and difficult to identify. Sometimes the fact that a community is suffering from an environmental malady may be very obvious even to the casual observer.

Action to restore the vigor and health of an environmentally impaired community depends on clear understandings of the nature of the impairment, the roots of the environmental problem, and the potentiality to take effective action. The purpose of this book is to describe selected efforts to diagnose and treat environmental maladies of various sorts in the community. We hope thereby to increase the competence and optimism of students and to provide as well an introduction to the scientific and social processes which are involved in community diagnosis and therapy of environmental problems. Environmental epidemiology is a young discipline, with what appears to be little of tradition, but this is partly an illusion. Primitive men had to decide what was edible, where to build a home or find shelter, how to avoid the dangers of burns from fires as well as the dangers of smoke inhalation. Avoidance of fecal contamination of drinking water, of contagion, and of poisoning were priority decisions for early man. At the very dawn of medical science, these problems were carefully analyzed as we note in Hippocrates' "Airs, Waters, and Places".[1]

The rapid expansion of competence in diagnosis and treatment of individual disease has tended to overshadow the older and more traditional concern with community environmental problems. In this sense, the resumption of attention to this set of problems is the essence of conservatism — a return, as it were, to the preoccupations of earlier ages.

The experience of communities in finding safe and wholesome water and food supplies is preserved in a host of cultural practices, just as is the experience in preventing and treating illness and injury of individuals.

In one important respect the situation of an afflicted community and an afflicted individual differs; the individual is accorded social supports and assisted to return to health and full functioning. The community in need of economic, technical, or educational assistance in dealing with its problem usually has to depend mostly on its own resources if the process of impairment is to be reversed. Only through the political process is it possible for communities with environmental problems to mobilize the help which many of them need and deserve. The political process itself is dependent on a clear demonstration of the nature of the problem, its causes, and the damages it is causing. Such a circumstance makes it all that much more important to learn how community problems were defined, and what the consequences of better awareness was in selected examples. That is one of the goals of the authors in writing this book. It is our hope and belief that by studying the problems of a few selected communities, and examining the consequences of the investigations carried out in such communities, the important relationship of the political process and the treatment of community environmental disease will be better understood.

Résumé of Section I.

- There are sick communities as well as sick persons.
- Just as the large number of diagnoses can be classified for persons, the smaller number of ways a community can be impaired can be classified.
- Environmental problems are an important class of disturbances in a community's health.
- Despite "environmental health" being a recently coined phrase, the problems, activities, and management strategies have a distinguished history.
- Social support systems are important for the sick person, and for the sick community political processes play the same role.
- Factual strategies are critical in mobilizing political mechanisms for the environmentally impaired community.
- This book seeks to provide the student with familiarity with such strategies as have been useful in a number of communities.

II. FUNDAMENTAL CONCEPTS OF ENVIRONMENTAL HEALTH AND THE SPECTRUM OF HEALTH

Analogously to the health status of an individual, there is a broad range of various indicators of health in the community, ranging from excess deaths, increases in illness, aggravation of the health status of those already ill, production of symptoms, impairment of function, and annoyance reactions. *All of these reactions or indicators in a community may have upward or downward trends and cycles either with or without these being related to environmental exposures.*

It is a common mistake of beginners to think in terms primarily of excess deaths, in part because of the lack of ambiguity of death. A moment's reflection will suggest that environmental exposures with excess mortality will usually have to be quite severe to produce excess mortality, and that there must be many more events for which the exposures are less and the effects likewise more subtle. To overlook these more frequent and more modest events would be a serious omission. It can furthermore be shown that in order to detect an increase in the number of deaths per day, due to an environmental exposure, either the exposure must be very intense or the population very large. So although excess mortality from environmental exposures is always to be considered and looked for, it will usually not be an adequate guide to identification of problems in most communities. It follows that no assurance of lack of environmental impact on health can be made merely because there is no evidence for excess mortality.

There is, however, a special set of circumstances in which mortality analysis is critical. These include mortality from causes for which the environmental exposure has some specific link. In the examples given in the next section, we consider excess deaths from cholera in relation to water pollution, deaths associated with an industrial "spill", as well as excess deaths among persons hospitalized with acute myocardial infarction (heart attacks) for which there was reason to think that the exposures to carbon monoxide might be unusually hazardous. Excess deaths among elderly persons are unusually important in the study of effects of heat waves. We assume that this is related in some way to the impaired adaptive mechanisms to any type of stress that is a characteristic of the elderly and ill sector of the population. It follows that excess numbers of deaths among the elderly would be a useful indicator of any environmental stressor; indeed in some locations, such monitoring already has been carried out as a guide to possible hazards to the general community from air pollution. The system did not prove to be useful in giving evidence of excess mortality among the elderly during high pollution periods in Los Angeles.

The most notable event detected by this system is shown in Figures 1 and 2. A heat wave both preceded and followed several days of ozone alerts (days in which the hourly average value exceeded 0.5 ppm). Excess deaths were related to the high temperature, but when the temperature dropped so did the mortality, even though the ozone alerts continued.[2] (See Chapter 6 for further discussion of heat wave experience in Los Angeles.)

The monitoring in Los Angeles was of deaths among persons residing in nursing homes; such persons are usually both elderly and with a variety of chronic illnesses and disabilities. It was thought that these people might be most vulnerable to any unfavorable change in environment. Indeed this monitoring program did detect significant increases in mortality associated with heat waves and during outbreaks of influenza, but did not identify any episode of excess mortality due to the notorious smog of Los Angeles.

By contrast, in London on several occasions, excess mortality has been documented, but the winter of 1962-63 was the last one with appreciable changes either in daily mortality or in hospital admissions that could be related to air pollution.[3]

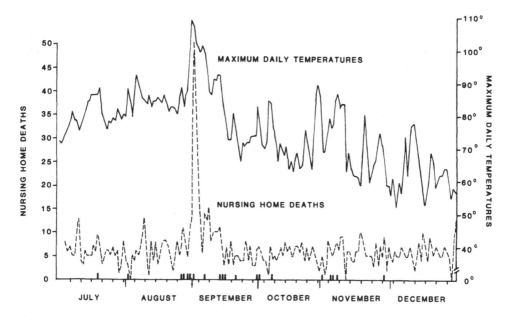

FIGURE 1. Relationship of heat wave and smog to deaths among nursing home residents in Los Angeles, 1955. The bars indicate days for which oxidant (ozone) levels exceeded 0.5 ppm for 1 hr. (From Goldsmith, J. R. and Breslow, L., *J. Air Pollut. Cont. Assoc.*, 9, 129, 1959. With permission.)

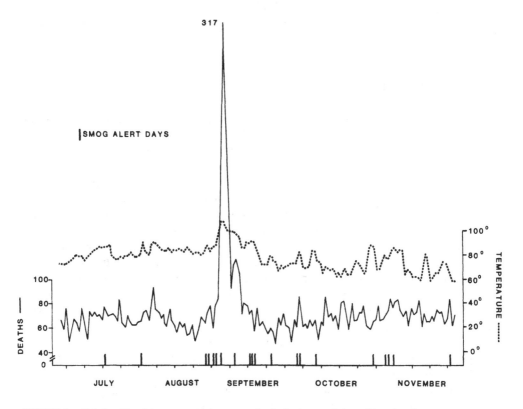

FIGURE 2. Relationship of heat wave and smog to deaths in the population of Los Angeles county, over age 65, in 1955 during the same period as in Figure 1, with the pollution levels as indicated. Note that the temperature scale is different, but the actual temperatures are identical to those in Figure 1. (From Goldsmith, J. R. and Breslow, L., *J. Air Pollut. Control Assoc.*, 9, 129, 1959. With permission.)

A. Definitions

For looking at data for the same population over a period of adjacent days or weeks the examination of *numbers* of deaths is useful and informative because approximately the same numbers of persons and the same sorts of persons are at risk. However when we wish to compare mortality for different populations, or even for the same location across a number of years, it is essential that we take into account the possible differences in the size and character of the population. This is most conveniently done by calculating the *mortality rate*, which is the number of deaths divided by the number of persons at risk. Since such a procedure usually results in a quite small figure, it is customary to represent the rate as per 100,000 persons.

Different cities and the same city over a span of years may have a population which contains different proportions of elderly persons, and since elderly persons may be expected to have a higher risk of dying than younger ones, we usually have to take into account as well the age of the population of a location when we compare its mortality rate with that of another location. We can arithmetically *adjust* the mortality rates for such a comparison in order to treat the data as though we were comparing populations of equivalent distributions by age. Such rates are then called *age-adjusted* rates. Alternatively we may want to examine the mortality rates for a specific age group or set of age groups. Such *age-specific* rates are of course the numbers of deaths divided by the numbers of persons in the specific age group (multiplied as above by 100,000 for conveniences in expression). Finally, we may compare *cause-specific mortality rates,* which is the number of deaths attributed to a given cause, divided by the population.

The *case-fatality rate* is the number of deaths from a given cause divided by the number of persons ill with that condition. The *attack rate* is the proportion of persons who become ill divided by the number who were "at risk" or were exposed to an agent likely to make them ill.

B. Some Fundamental Concepts and Problems

Now that some of the terms used to describe health reactions and to compare them from place to place are defined, we can return to some fundamental concepts of environmental health.

If we examine the age-adjusted cause-specific rates for certain sites of cancer, such as cancer of the liver, lung, or stomach, we find that the rates differ very widely. It is often assumed that the excess above that of the lowest place is due to "environmental factors". Such assumptions should be considered skeptically until such a factor or factors can be specified and other possibilities such as genetic factors, exposures to infectious agents, and the general conditions of medical care are excluded. There is, after all, no health benefit from being told that you or your community have an excess cancer risk due to "environment" unless some modifiable factor is obvious or specified. So although variation in mortality and morbidity (sickness) rates from place to place and time to time are commonly said to be due to environmental factors, unless the factors are obvious or specified, such attributions are of no immediate benefit. Only when such attributions lead to a search for an environmental factor is there likely to be some benefit.

If increased mortality or illness (morbidity) rates are related to environmental exposures in a given community, it follows that a somewhat greater frequency of effects of lesser severity are likely; they should be looked for, since such milder reactions may be the more readily prevented and monitoring of them may provide a guide for the adequacy of environmental management. Thus, if it is established that an environmental exposure causes one type and level of health effect, others should be looked for, especially milder forms of reaction on the spectrum of health effects.

Environmental exposures do not usually cause a specific type of illness or impairment which is not also caused by other factors. That is to say, most environmental health effects are *nonspecific*. Bronchitis, asthma, cancer, heart disease, and impaired lung function may have many causes and if we wish to attribute them to a given environmental exposure a causal role, we shall have to exclude or somehow adjust for the effects of all the other factors which may be relevant.

C. Estimating Environmental Exposures

Just as rates are important for measuring health status, so a sort of a ratio is used to estimate environmental exposure to chemical or physical agents. We measure some amount of the chemical in a given amount of water or air and give the result in terms of a *concentration ratio* such as parts per million (ppm) or micrograms or milligrams per cubic meter. Other environmental exposures are often expressed in standard units such as atmospheric pressure in millibars, temperature in degrees Celsius, wind speed in meters per second, etc.

Physical or chemical measurements must be evaluated as to their *specificity* and their *precision*. Specificity means that the measurement is what it really is intended to be, rather than a reflection of some other material which causes the instrument to give a positive reading.

Since the amounts often measured are so very tiny we need to know how accurately the measurement is made and that is one of the meanings of the *precision* of a measurement. We may also use the term to indicate the minimal detectable amount of the estimated variability of a measurement.

Most environmental measurement systems are set up at a given location, convenient to electricity and water supplies and where the equipment is secure from vandalism. It is then usually assumed that the measurements obtained are suitable for characterizing the exposures to persons in the vicinity. In fact most air pollution instruments sample with an acceptable level of validity and reliability the outdoor air at the site where they are located. Since most humans spend more of their time inside a building than outside, such measurements are of limited validity as an estimate of the total exposures of a population. Scientists and engineers are just beginning to measure indoor pollution levels and to attach portable samplers to a person in order to obtain a "Personal Exposure Estimate". In fact we must accept that estimating population exposures is difficult and that *fixed site sampling is at best a way of approximating the different exposures* of populations living and working adjacent to such sites. Similarly, measurements of pollutants in drinking water supplies are of limited validity as an estimate of the exposures of a population drinking this water to the pollutants measured, because not all of the fluid intake of individuals in the population comes from the community drinking water supply.

We are able to measure some materials with a greater validity, precision, and convenience than others and it follows quite naturally that most of our measurements are of such materials, even though they may not truly be the materials of most importance to community health reactions. Thus, we can easily and validly measure the sulfur dioxide concentration in the air, but what may be more important to our health is the sulfuric acid aerosol that is the result of the oxidation and hydration of sulfur dioxide. Measuring it is more difficult and is therefore rarely done. So environmental measurements, even if valid and relatively precise may nevertheless also be nonspecific regarding their relevance to the health effects associated with elevated levels.

There are important and often specific time relationships between exposures to environmental agents and the manifestation of the resulting health effects. Even the acute effects of pollution on mortality are not instantaneous, but may be delayed 12 to 36 hr. Other effects may have even longer *latent periods*.

For example, exposures to lead at levels common in pollution take weeks to months to build up in the body to the point that some effects can be detected. Exposures to carcinogenic agents (materials which can cause cancer) such as asbestos and arsenic may have latent periods of 30 years or more.

Résumé of Section II.

- A spectrum of health reactions is relevant to environmental exposures; it ranges from excess deaths, excess illness (morbidity), aggravation of pre-existing disease, production of symptoms, impairment of function, to annoyance reactions.
- The trends and cycles which can be observed in health variables may or may not be due to environmental exposures.
- Excess deaths, although commonly and properly considered, are not likely to be the most sensitive indicator of environmental harzard.
- Once a given effect has been observed to be related to a given environmental exposure, it is likely that effects of a more "minor" type can also be found.
- Mortality analysis is likely to be especially useful for sensitive groups such as the elderly or for conditions specifically related to the exposures.
- A variety of rates for mortality and morbidity are in common use; they include (crude) mortality rates, age-specific and age-adjusted mortality rates, cause-specific mortality rates, case fatality rates, and attack rates.
- Environmental factors are often offered as "explanations" for temporal (referring to time) and spatial (referring to locational) variations in mortality or morbidity, but unless the factors are specified, such statements are not usually helpful except as an indicator of the need for further study.
- Environmental exposures often have nonspecific effects which could also be caused by other variables.
- Estimations of environmental exposures are for chemical agents usually in terms of a concentration ratio. For physical agents standard units are used. Measurements need to be evaluated as to their specificity and precision.
- Most environmental measurement systems are at a fixed location, whereas most of us move from place to place. Indoor pollutant measurements and personal sampling are giving us a more valid picture of the doses to which real populations are being exposed.
- Many pollutants which are being measured are not themselves the agents likely to be causing reactions; the pollutants we often measure may be substitutes (or surrogates) for the agents which are really harmful.
- There is usually a delay between the exposure to a pollutant and the manifestation of a health effect. This latency period may be as short as a few hours or in the case of cancer as long as 30 years.

III. HOW SCIENTIFIC APPROACHES DIFFER

The use of numbers within a framework of logic in an objective manner is what distinguishes scientific from nonscientific activity. In the section which follows we are going to pay particular emphasis on the use of numbers. The framework of logic is at least as important. Goldstein and Goldstein have provided an excellent introduction to this in their book *How We Know: An Introduction to the Methods of Scientific Research* (Plenum Press, 1978).

There are two kinds of numbers we commonly use, counting numbers and measurement numbers. In determining the standings of a ball team we tabulate the numbers of games each team has won and lost and calculate the percentage of games won. We determine the best team by using these counting numbers. But in a track meet, we measure the elapsed time of each runner, the maximum height of the jumper's leap, and the gold medal goes to the individual with the "best" set of measurements.

Table 1
RESPIRATORY ADMISSIONS TO CLINIC BY DAY ACCORDING TO POLLUTION

Day	Pollution alert (+ or −)	# of Admissions
1	−	23
2	−	30
3	+	32
4	+	28
5	−	33
6	−	27
7	−	18
8	+	26
9	−	25
10	+	26
11	+	29
12	+	24
13	−	35
14	−	19
Sum	+ = 6, − = 8	375
Average		26.79

In science, although we make many comparisons, there is as it were a secret competitor, the operations of chance. We commonly are asking the question, "Could this result occur due to chance variation in the conditions of the test or study?" Statistics is the scientific discipline which helps us decide if the observations we have made are outside the range of what could be expected due to random or chance effects under the circumstances of the study. It is because the statistical treatment of measurement and counting numbers differs that we have learned to make the distinction.

Table 1 for example shows the numbers of persons admitted to a clinic with respiratory problems on days with (+) and without (−) pollution alerts during a two week period.

If we sum the numbers of admissions in the artificial data set for the days with pollution alerts we find on those 6 days the number of admissions was 165, for an average of 27.5 a day and that on the 8 days with no alert, there were 210 admissions or an average of 26.25. By the rules of sport, we would declare that the pollution alert days had "won"; the rules of science impose a logical precondition that "all other things are equal", by which we mean that there is nothing about some of the days which might make both admissions and the likelihood of pollution differ on those days from the other days.

When we look carefully at the table we note that the two lowest days are 7 and 14, and that both of them were days without a pollution alert. In all likelihood, days 7 and 14 are Sundays, and it is usual to find that sickness onsets seen at clinics are fewer on Sundays. It happens also that the sources of pollution such as factories and automobiles emit less on weekends, so we would expect that fewer Sundays would be air pollution alert days than would be the case for other days of the week. If these 2 days are excluded from the set of nonalert days, then the total for the remaining 6 days is 173 admissions, for an average of 28.83 admissions per day, a number slightly higher than the average for pollution alert days. If what we have been testing is the hypothesis that there are more respiratory admissions on days with pollution alerts, we would "reject" the hypothesis on the basis of these data.

This sample of 14 days contains 6 nonalert weekdays. If we assume that they are a representative sample of all weekdays without pollution alerts, we can use statistical methods

Table 2
RELATIONSHIP BETWEEN HOURLY MAXIMUM POLLUTANT MEASUREMENTS AND TIMES FOR COMPLETION OF A FOOT RACE OR A SWIMMING RACE

Day	Pollutant con- centration (pphm)	Foot race time	Swimming race time
1	16.6	23.9	63.7
2	24.3	21.8	63.0
3	15.3	19.6	61.8
4	6.1	19.1	60.1
5	10.0	18.3	61.8
6	21.1	19.8	62.5
7	17.7	21.2	60.8
8	26.3	22.9	60.3
9	28.5	24.0	62.7
10	4.8	20.3	62.2
Average	17.07	21.09	61.89

to determine how many admissions would be necessary in order to be outside the bounds of chance. We usually accept the convention that if an occurrence would only happen 1 time in 20, it is outside the bounds of chance.

We can use the standard deviation (SD) as the basis for this calculation. The SD is an index of the variability of the measurements or observations around their mean; it measures the extent to which the observations are dispersed or clustered closely together. Numerically it is the square root of the mean of the squared deviations of each observation about the mean value of the set. The SD of the numbers of admissions on those 6 days is 4.67. An observation greater (or less) than the mean plus about two SDs has a probability of about 1 in 20 of occurring as a result of chance, so twice 4.67 + 28.83 is 38.16. We can then say that if we observed at least 39 admissions, that would be a "significant" increase in the number of admissions; conversely if on a weekday we observed 19 or fewer admissions computed from $[28.83 - (2 \times 4.67)] = 19.49$, that would be a significantly low number.

Note that the chances are 1 in 20, or in decimal notation, $p = 0.05$, that a number of admissions outside of two SDs *in either direction* would be found. If in fact we are only interested in looking for deviation in one of the two directions, the probability is just half that or $p = 0.025$ for a value two SDs above (or below) the mean. Of course in the statistical treatment of counting numbers, no decimal fractions are possible for outcomes, so we have in the above examples cited the next higher whole number, 39, and the next lower whole number, 19, as the events which would just exceed two SDs from the mean in either direction.

There are many other statistical tests which are specific for counting numbers, and some of them occur in the examples which will be presented in the next section.

Let us now consider an example dealing with measurement numbers. As in Table 1, the numbers in Table 2 are artificial, but they are intended to represent the measurements of the highest average for a 1-hr period of a pollution measurement which we assume is valid for the day and location, and the time for a given student athlete or the mean for a team to complete either a foot race or a swimming race.

What does this table tell us about whether there is or is not some relationship between the pollution levels and athletic performance? First, we assume that all days are weekdays, so that days are not dissimilar in that respect. There are two other ways in which days may

Table 3
MEANS AND SDs OF THE VALUES
FOR THE 5 DAYS WITH THE
HIGHEST POLLUTION COMPARED
TO THOSE FOR THE 5 DAYS WITH
THE LOWEST POLLUTION FROM
TABLE 2

	Pollution levels	Foot race times	Swimming times
Highest			
Means	23.58	21.94	61.86
SD	4.27	1.61	1.22
Lowest			
Means	10.56	20.24	61.92
SD	5.30	2.17	1.28

be noncomparable. First, for student athletes, we expect that, other things being equal, there is some learning and training effects, such that earlier days have slower times than later ones. Just looking at the data suggests that this is more the case for the foot races than it is for the swimming times. Still the examination of the table doesn't tell us much. Following the criteria used above in the discussion concerning Table 1, we might look at the means and SDs of the 5 days with the highest pollution levels and the 5 lowest days, and then test whether the highest days differed by more than chance (that is by more than two SDs from the mean for the lowest days) Table 3 shows what these values look like.

Except for the pollution levels, which we have arranged, there are no differences greater than two SDs, and only the foot race data show that the better times are for the days with less pollution.

If we graph the data we can see that there does seem to be a relationship between pollution levels and slower races. The numerical index of the closeness of the relationship of two such variables is called the correlation coefficient. The correlation coefficient is a number between −1 and +1. A coefficient of −1 indicates that variation in the two measurements or observations is perfectly matched, but that they vary in opposite directions. A coefficient of +1 signifies that a variation in the first measurement is perfectly matched by variation in the second measurement and in the same direction. A coefficient of zero (0) signifies that there is no relationship between variation in the two sets of values.

When we calculate the correlation coefficient for these data, we obtain for the foot race data a correlation of 0.69, whereas for the swimming data it is only 0.23. As large a correlation of 0.69 with ten observations is unlikely to occur by chance, so we can say that there does seem to be a statistically significant relationship between pollution levels and running of foot races, if the effect of training and learning is ignored and "all other variables are equal". For further discussion of a real data set dealing with possible effects of pollution on running times see Section VII. (Students with a background in statistics may want to compute the standard error of the foot race difference in Table 3 or perform a t test, or see if adjustment for learning affects the results.)

In these examples, we have *counted or measured* the numbers of admissions and the times it took an athlete to complete a standardized race. These measurements are for a *sample* of all possible days or races. We have used the SD of the sample as a basis for *estimating* the amount of variation to expect from another sample of days or races.

If we want to avoid having races on days for which environmental conditions might affect the outcomes, we will need to determine either how much pollution it takes to have such an effect, or alternatively we may repeatedly measure the times of a group of athletes in order to see if on certain types of days, their times are slower. This is called *monitoring*, and means a systematically repeated set of observations for the purpose of determining what course of protective action to take. We could have monitored the environmental conditions, in which case we conduct environmental monitoring, or we could monitor the health reactions, in which case we call it health monitoring; if the health monitoring is based on a population sample we call it epidemiological monitoring.

Résumé of Section III.

- Scientific method is based on using either counting or measuring numbers within a framework of rules of logic using objective criteria.
- Small differences in data sets may be due to chance variation and we use a variety of statistical criteria to determine if a given result is unlikely to have been due to random or chance variation.
- In comparing measurements or counts representing phenomena for a given day or period, we have to ask and satifactorily answer the question: "Are all other attributes of these days or periods equal or comparable?"
- In doing statistical tests, we need to decide if we are testing for a value beyond the expected levels of chance variation in either direction, or whether we are only interested in variation in one direction.
- We test hypotheses to see if there is or is not a difference between the results that are statistically significant according to whether the difference exceeds the bounds of what is likely to occur as a result of chance variation.
- As a first step in determining whether a difference is in the direction of a hypothesized difference, we look to see if there is any difference at all in the direction which would support the hypothesis.
- We are often dealing with samples of a large set of events for which the sample gives us estimates both of the "central tendency" or mean (average), and the "dispersion", estimated by the standard deviation (SD) of the sample.
- Monitoring of either environmental or health reactions is often useful as a guide for action. We use the occurrence of deviant or unfavorable experience as a triggering mechanism for further protective action or additional measurements.

IV. TYPES OF STUDY DESIGN FOR COMMUNITY STUDIES

There is a natural sequence of approaches to measuring the possible health effect of environmental exposures in a community. The initial suspicion or inquiry may arise because there seems to be an usual number of asthma patients requesting refills of their prescriptions on days with the wind blowing from a given direction; or patients with heart failure complain that they are more short of breath on days when they are bothered by odor from a local mill or plant; in the event of a major heat wave or industrial spill, the question isn't whether some hazard has occurred, but how bad it was for the health of the community and its members. But there must always be a question rasied by someone, and of something.

The first step is to see what background data are available on pollutant emissions or levels, as well as the possible health reactions. For example we want to know the usual number of prescriptions per day or week and what proportion of heart disease patients usually complain of shortness of breath, bearing in mind the importance of other variables, such as day of week, age of the patients, experience with the disease, as well as possibly biassing factors such as attitudes of the community and industries or physicians. If, from this sort of review, the suspicion of a problem seems justified, then the next step must be decided upon; usually that is to obtain further data, but occasionally the mere suspicion is sufficient to take protective action, based on the general fund of knowledge and relative simplicity of providing protection.

The next step is usually what is called a *retrospective study,* in which what happened previously is documented and analyzed, in order to confirm or refute the suspicion or to determine at what level of exposure effects might be found. Retrospective studies may be made by examining records or reports, or by interviewing those involved and potentially affected. The classic form of such a study is what is called a *"case-control" study.* For example of all the heart failure patients in the clinic, the cases are those who had complaints of shortness of breath, and the controls are those who did not. We then will want to try to find out if there are any ways in which those who did have such a complaint differ from those who did not, particularly with reference to the occurrence of the complaint on days with a noticeable odor in the air. (Some of those who were not bothered may have been indoors all day, have an air conditioning unit with a good filter, or are relatively inactive, for example.) Suppose that there were 200 patients cared for at the clinic who had had heart failure, and we decide that we want to send a questionnaire to each one; what questions shall we ask? True, we want to know name, address, activity patterns during the past week, type of housing, whether or not air conditioning is installed and what type of filter is used, and of course whether or not they had more shortness of breath on some days than others. *What we don't do is to ask the direct, leading question: Did you have more shortness of breath on days when there was a detectable odor from the plant?* This contrasts sharply with the usual type of medical diagnostic interview, in which leading questions are the rule.

We may, for example, ask "On which days during the past week were you more bothered by shortness of breath?" "What was there about such days which you think may account for this?" Or alternatively, we could ask "Have you been bothered by odor from the plant during the past 2 weeks?" and "How were you bothered?" Or we can ask the even more general question, "During the past 2 weeks, were there any days on which you did not feel well, or on which you felt worse than usual? In what way?" In Chapter 11 we illustrate some results of a community odor survey.

We may be able to divide the area which is served by the clinic in such a way that persons living in one area are more likely to be exposed to pollution from the plant than are persons living in another area. In this event we can compare the questionnaire responses for the two areas, with respect to the number who had shortness of breath when the wind was from the plant or when they noticed an odor. In this event we are testing a compound hypothesis, that persons living in the area most likely to be exposed to the emissions from the plant are most likely to have disease aggravation (that is shortness of breath), and that they have this effect at the time when the wind is from the direction of the plant. We call this a *"Temporo-spatial hypothesis",* because it involves both a criterion involving time (temporo-) and one involving location (-spatial).

This type of study involves *all* of the persons thought to be at risk of the effect we are studying (shortness of breath among patients of a heart disease clinic). Should we want to know how many persons in the entire community were bothered by odor from the plant? It would be inefficient in most communities to try to interview everybody. Instead, we usually draw a *probability sample* of the residents, often stratified according to the proximity of the plant and therefore the likelihood that they would be bothered. (Chapter 11 illustrates these principles.)

The second major type of study is *prospective,* that is it is intended to obtain information which will occur in the future about the relationship of pollution exposures and health or community reactions. Since such studies are usually more expensive and the results are uncertain due to all the unpredictable things which may happen in the future, it is usual to precede a prospective study by a retrospective study either of data collected for other purposes or of survey data. A common type of prospective study is a *cohort or panel study.* The word "cohort" means a group of persons with some common attribute, such as age, place of residence or place of employment, who are to have their health status followed forward

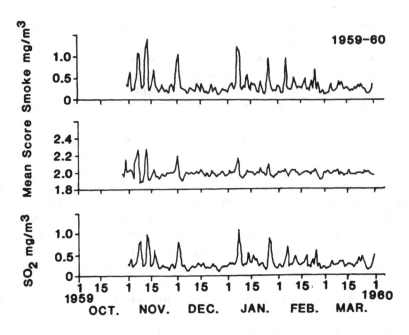

FIGURE 3.   The association of the mean degree of illness in a panel of patients with chronic bronchitis in London during the winter of 1959-60, middle panel, and the fluctuations in daily sulfur dioxide levels, lower panel, and smoke, upper panel. Large fluctuations in pollution early in the season are associated with worsening of chronic bronchitis, but smaller fluctuations and those late in the season do not appear to have much effect. (From Biersteker, *Environ. Res.*, 11, 296, 1976. With permission.)

in time. *Panel studies* are usually of persons with a given health status or problem from whom we have made an arrangement to collect information periodically on their health experiences. We might do, for example a cohort study of all children born during September, whose health we follow during the next winter in order to see if indoor pollution leads to increased frequencies of respiratory diseases for which they need medical care. Usually a cohort is a population of healthy individuals at the outset of the study period. A panel study is, by contrast usually a group of persons with some long-term health problem, such as persons with chronic bronchitis or asthma, from whom we want to obtain at regular intervals answers to such questions as: "Was your condition today better, worse, or about the same as it usually is?" Since subject cooperation is critical for the prospective study of a panel, we have learned by experience that cooperation of a panel begins to decline after about 3 or more months, so it is not usually possible to extend a panel study for a long period of time. A very significant set of panel studies was initiated by Prof. P. J. Lawther to follow the effect of black suspended particulate matter and sulfur dioxide on the health of chronic bronchitic patients in London.[5] Figure 3 shows the response to a panel of bronchitic subjects in London to pollution during the winter of 1955-1956. The "Clean Air Act" in England led to a great reduction of pollution in the early 1960s and as such panels were monitored for successive winters, the response became less and less and is no longer detectable.

In the U.S., panel studies are used to monitor the occurrence of eye irritation, and for the study of asthma (see Chapters 9 and 10). An important potential for panel studies is in classifying persons according to their apparent reactivity when air pollutants vary as they naturally do, and the determination of what attributes account for the variation in reactivity.

A third major classification of studies is *cross-sectional;* that is studies of a sample of a population at a given time, in order to determine what their current health status is as well as their health and exposure history. Such studies are of most help for evaluation of relatively long-term effects, such as persistent cough or diminished lung function.

A common sense balance needs to be maintained as to just how much data may be required in order to take protective action. Where and when the cost of such action is modest and little disruption of ongoing activities are involved, a relatively small amount of data may be needed to provide a basis for taking action. On the other hand, if major changes in industrial, transportation, or energy use are likely to be needed in order to abate an incipient problem, or the costs are high, a very large scale and sytematic effort may be justified.

Résumé of Section IV.

- From the point of view of time there are three classes of studies, the retrospective, the prospective, and the cross-sectional.
- The usual sequence of events is that some observation leads to a suspicion of an effect, for which a retrospective study either of records or using interviews provides supplemental information. If further confirmation is needed, either prospective or cross-sectional studies may be done.
- Interview questions should not generally be leading questions, in order to avoid telling the respondent what you the investigator are especially interested in; the answers to such questions are not likely to be objective.
- Prospective studies may be of panels, usually of persons with a long-term illness, or of a cohort; in panel studies we usually require periodic reports, so the duration of panel studies is usually limited to a few months.
- Studies may be of all the persons exposed or of a probability sample in which each person's likelihood of being sampled is identical; some sampling schemes may be stratified by various attributes, in which case the uniformity of sampling probability applies to the individuals in the strata.
- Studies in which the hypothesis to be tested involves variation by both time and location are called "temporo-spatial strategies".
- There is no fixed relationship between the amount and type of data needed to justify protective action, but if the action has a relatively low cost, a well-documented suspicion may be sufficient to justify it.

V. HOW TO OBTAIN AND EVALUATE ALREADY AVAILABLE DATA

Fortunately, a number of the data sets in which we may be interested are accumulated by ongoing activities of a government, school, clinic, or other institution. It is important to be able to take advantage of such data sets. A major set consists of the "vital statistics" which are accumulated by the vital records agency, usually of the health department. These include all births, deaths, marriages, and divorces. They are usually collected by county and state health departments, and checked and tabulated by the U.S. National Center for Health Statistics.[5] Although there may be charges for obtaining tabulations, and confidentiality protects access to information for a specific person or family, the data tabulations are available for public scrutiny. There may be considerable difficulty in finding who has access to which files, and some delay in producing the information one wants, but the inherent validity of the data is usually good, and if they can be accessible, the analysis is straightforward and the interpretations usually influential.

However as indicated earlier, mortality data usually are not adequate as a basis for detecting the most common and milder impacts of pollution in the community, and data on morbidity are not as well standardized, not available from obvious sources, and may have such substantial biases which are not easily identifed that their use may lead to erroneous interpretations. One of the best standardized types of morbidity data is cancer registries; their only disadvantage is that cancer, although often feared is rarely if ever a result of exposures of general populations to pollution. National Health Survey data is another source of morbidity data, but the data usually does not represent any local area, since it is designed to reflect national conditions. Locally, hospital admissions data may be useful, but confidentiality often restricts access to it in ways which render it essentially unavailable. However, there is no confidentiality barrier to a hospital making available data, say, on the number of respiratory admissions it had on a weekly or daily basis, the number of requests it had for emergency room treatment of asthma, or the frequency of burns, or motor vehicle injuries among children. Some health insurance or health care systems, including organizations such as the Health Insurance Plan of New York, Group Health Cooperative of Seattle, or the predominantly West Coast, Kaiser Health Plans have data tapes on hospital admissions or new onsets of some illnesses for their more or less stable and locally identifiable populations, but only under unusual circumstances have such data sets been made available for analysis, and when this did occur, it was usually to well qualified academic or professional colleagues. It is possible that membership interest among such groups may stimulate a greater potential for such evaluations.

A local school system keeps records of attendance as do local businesses and governmental agencies. If one assumes that absence is related somehow to environmental exposures, the analysis of such data may be useful. Some of such data include reasons for absence, and if that is so, the data may be unusually helpful. If the local health officer, school nurse, school principal, or head of a parent-teacher organization is part of a community-wide committee, then availability of such data becomes more probable. There are many other factors which influence both school and work absence, and although a number of studies have examined such data, it is usually only useful for getting leads, and not for drawing conclusions. Such data are more likely to be useful if one nurse, clinic, or physician obtains it and if the medical basis is clearly specified.

Physician office or clinic records may be a useful source of information, provided no information is disclosed concerning an individual patient. If the physician is interested in community problems or is a member of the community-wide committee, access may be that much more likely. Sometimes the physician's interest may be such as to bias the data, since physicians are trained to ask leading questions in the clinical setting. Unfortunately, many physician's office records are inconsistent and poorly organized, in contrast to the somewhat better state of medical records in most hospitals.

Public agencies, such as police, pollution control agencies, and health departments receive a great deal of information from the public as to complaints, injury, and other events. Summary data as to the frequency of such events is often of considerable value and once again, as long as it does not refer to a specific individual should be available for the use by scientists and citizen groups.

Recently, "Right to Know" laws have been enacted which affirm that workers and persons resident in the vicinity of industrial plants have a right to know to what materials they are exposed in oder to be able to protect their health. Supported by labor unions and consumers groups, the demand of the public to know the nature of the hazards they must deal with is an important resource in conducting community studies. Such laws although originally directed against industries which claimed that disclosure of exposure risks would require them to give up "trade secrets", may also be a means of requiring disclosure of exposure information from regulatory agencies.

Newspapers receive, process, and store a great deal of information as to industrial developments, police and fire reports, meetings, and community problems. Local newspapers deserve scrutiny for the leads they may give, although the data likely to be found may sometimes need to be validated.

Local pollution control agencies can be expected to provide data on pollution monitoring for which they are responsible for use by scientific and citizen organizations; sometimes state or federal agencies are better organized to provide local information and to provide reference information needed to interpret local information.

Sometimes there is a cost or delay in providing the information. If in seeking some information you think may be useful and even urgent, you are told that it cannot be made available until after a few months, or it is only available if funds are provided for clerical support or statistical treatment, don't interpret this as a stall or rejection, even though it may be. It may also be true that even the most sympathetic agency, clerk, or office cannot take on unbudgeted tasks, and that it really does take some time to form a new set of tables from otherwise accessible data. It always makes the process easier and more likely to succeed if there is a budget, and the community group has a planning process which allows for the needed time to organize the data collection tasks which may be needed.

From this discussion, any alert community-oriented person or group can identify special potentials which may exist in their community and obtain some suggestions as to ways to make the data they may need more likely to become available at a reasonable time and at reasonable cost.

Résumé of Section V.

- Official vital statistics are usually available from health departments on a local, state, or national level; particular interest attaches to birth and death records. However death rates are usually not very sensitive to offensive and objectionable levels of pollution. Delay and cost may impair their usefulness, but they have a high validity, and if possible should be included in any systematic report.
- There are some official sources for sickness or morbidity data, but they are not likely to be of much practical value.
- Other morbidity data are often collected by health insurance organizations or health maintenance organizations, but they are not usually available to community organizations.
- Although hospitals and physicians collect a great deal of potentially useful data, individual items are confidential and only summarized data can be made available, but that is often just what is wanted; however some time and costs are likely to be required.
- "Right to Know" laws reflect a growing awareness of the need of individuals to have and to use information as to the hazards in the community and workplace in order to protect themselves from such hazards. Such laws can also help obtain data needed for planning and evaluating community environmental hazards.
- Newspapers may be a rich source of information, but it is often in need of careful sifting before it can be relied on. Assistance in obtaining access to data can often be assured if key officials with access to potential sources of data are placed on a citizen action committee.
- Local governmental agencies such as pollution control agencies and health agencies have an obligation to provide information about local problems, but there may be budgetary or time constraints of what they can do.

VI. HOW TO ADAPT TO REALITY AND LIMITED RESOURCES

The first step in obtaining the resources needed for a community study is to know what the right question is and to whom to address it. If you are concerned about the health impact of a factory, smelter, or pulp mill, you will need to know who both the managers and owners are, what local regulations are, what the record of the plant is with respect to compliance with these regulations, and if the regulations are not being complied with, what are the legal remedies. On the basis of that information, you may decide to go to the plant manager, the local medical association, the local pollution control office, the district attorney, the parent teachers organization, or lung association. Not all of these addresses themselves can offer you funds, but each of them has a stake in getting the answer, and letters of support may make a big difference when you do come to ask for support.

The question is usually a variant on "What does this stuff do to our health or that of our children?" This can be followed or preceded by "If it smells this bad..." or "If it is this serious a problem for workers...." or "If this is a potent chemical, why should it be spread all over our neighborhood..?"

At the same time, you should be prepared to know what information you can find out without getting any additional support, because this information is the principal lever you may have for getting help. You should expect to get either information or backing or funds, with the former being much more likely than the latter. Be ready to use your local college, university, and community associations by understanding their own missions and putting the community's environmental problem in terms which they recognize as either a community or teaching responsibility. This process of looking for support will itself be an educational one for all of you.

You must know at the outset that substantial funds for studies usually come from the federal government; you should also know that usually these funds go to well-established organizations for well-studied problems, but one of the local colleges or universities may be such an organization, and their involvement may be of great assistance for extending their contract on the basis of the question you pose and the community contacts which you bring with the questions you are raising. The likelihood of state or local governmental support often depends on the professional or occupational background of one or more legislators or commissioners.

Of course if you, yourself are a local governmental official, or faculty or staff member of a local college or university you have a different role.

Finally, the local newspaper may or may not be interested in the question you are continuing to ask, and don't forget how helpful a photo is of dust clouds, children having to restrict their play, or even lines at the local clinic.

This is an imperfect world; pollution is one of its imperfections. Somebody's indifference, ignorance, or greed may be largely responsible for the pollution problems in a community, but it is also possible that lack of perfect foresight allowed a problem to get out of hand or that new perspectives permit us to recognize a pollution problem which previously was obscure even to conscientious, well informed and unselfish community and industrial officials and managers. It is often very difficult to determine in advance who, if anyone is to blame, although it is a great temptation to find a scape-goat.

A long-term perspective is often the most suitable basis for resisting this temptation and the consideration that dealing with a community environmental health problem is an educational venture. Polluters, regulators, and citizens need to learn not only about the facts relating to exposures and possible effects, but also about a new set of priorities, which alone can lead to effective means of preventing health damage from pollution. It is true that there are many useful regulations on the books which can penalize a polluter, or require alterations in technology, but behind the usefulness of regulations lies the resources of the person who enforces them and the person to whom they are meant to apply. These people are the essential part of the critical path from the often negligent route of pollution to the ethic of environmental conservation. Try to understand their backgrounds, motivation, and potentials for playing a useful and constructive role in your community. On the other hand, being well informed, resolute, and resourceful is the best defense against evasion and digression.

The more convincing your identification with the community's interest is, the more effective you will be in obtaining support for the effort to study a problem and for disclosure and abatement of any problems which the facts point to.

On the other hand, although you may be able to define very clearly what information you want to have, you must expect that much of it either does not exist, is of poor quality or validity, or cannot be made available at a reasonable cost or within a reasonable period of time. You may also expect to get information which is not at all reliable or appropriate, and you should be capable of sifting the useless or misleading from the useful items of data; if necessary you may have to get technical help in order to do so. Nearly every study of environmental health problems in every community, including the ones which are included in the next section of the book represent some compromise or adaptive process with respect to the information which the investigators would have liked to have had.

One must adapt one's objectives to what is possible; often knowing what sources are available makes it possible to substitute a realistic plan for one that is not practical. Adequate control for all of the confounding variables is not always possible, but often a novel method of presentation or analysis does make it possible to draw a limited conclusion which is sufficient to start the process of correction, improvement, or prevention.

Limited resources are the rule, not the exception in this field. One way to stretch them is to involve the key providers of data in the planning and community action, and to encourage them to include the work you want to have done in their ongoing programs as a public service. Another is to have a thorough knowledge of what the data resources in your community may be. A first rate science writer for the local newspaper can be a powerful ally in making the case for the needed support both with respect to resources as well as the cooperation of officials in government and industry. In return, subject to other considerations, such a reporter may expect that your findings will be made available to him/her so that he/she can do a story on it.

Applications for support of studies on pollution effects may be submitted to a variety of funding sources, governmental or voluntary. In dealing with business or industries, remember that your study may be of great consequence to both their "bottom line" as well as to their public image and employee morale. Don't be reluctant to accept their financial support or that of their trade association; don't on the other hand be surprised if they show a desire to predetermine the outcome which some agencies and companies have a difficult time resisting. You may have to insist on objective procedures and open reporting of results, but in the long run you have no constructive alternative, so be prepared to be stubborn and expect that when the last page of your report is written and submitted, it may please no one or it may please everyone. Above all it should represent progress in understanding of the technical, social, political, and scientific aspects of the problem and the community which is affected by it.

Résumé of Section VI.

- If you want to get any study started or action taken on a community environmental health problem you must:
- Know your community
- Identify with your community
- Have the patience of Job
- Have the shrewdness of Machiavelli
- Have the integrity which Diogenes sought
- Pray
- It may also help to have studied the rest of this book
- Remember what Pasteur said: "Chance favors the person whose mind is prepared."

VII. SOME PRINCIPLES FOR THE INTERPRETATION OF NUMBERS

Much of the data from community studies of environmental health problems are in numerical form, and anyone who has ever looked at the monthly report of an air monitoring station or a water quality laboratory knows that many of the numbers are difficult to interpret. Even the terms may be obscure; maximum daily average — BOD (biological oxygen demand) — most probable number (of bacteria) are jargon.

In previous sections we have mentioned some statistical terms; standard deviation, correlation coefficient, and mean or average, to be specific. Although you should make sure that these common statistical criteria are known and useful to you, the first principle of numerical interpretation is not a statistical test, but graphical. Or as a statistical colleague has put it; " The criterion of interocular impact comes first; does it hit you between the eyes?"

Let us consider a set of real data on the performance of high school cross-country track teams in Los Angeles Basin from 1959 to 1964.[6] The data set consists of the times for all boys who ran in all home meets during these years. All meets began at 3 o'clock pm; the season runs from mid-September to mid-November. The number of boys who ran in all of the home meets varied from 11 to 30, and there were three to six home meets depending on the year. This is an example of using an existing data set, because the air pollution monitoring station was two miles from the school, and all of the data were available before the investigation began. The mean time (in seconds per mile) decreased over the course of the season as the boys became better trained and more experienced, from about 363 to 385 at the start of the season to 342 to 363 at the end of the season. Since the photochemical oxidant was usually greater in September than in November, a simple correlation analysis would be expected to show that oxidant was spuriously associated with slower running, because seconds per mile was high at the beginning of the season. It would be considered to be spurious because the time of the year was the true "cause" of both the decrease of oxidant levels and improvement in running times (Figure 4).

The scientists who wanted to study the possibly causal relationship of oxidant exposures to athletic performance got around this problem by converting the measurements (mean running times) to counts, namely the proportion of boys whose running times were greater on a given day's meet than they were in the preceding meet. Thus the analysis was made based on the percent of the team whose performance was poorer on a given day than on the meet immediately preceding. Table 4 shows the data.

Before looking at these numbers, a plausible set of hypotheses would be: (1) that because carbon monoxide specifically interferes with the transport of oxygen to the tissues of the body, there should be a close correlation between the carbon monoxide levels the hour before the race and the performance; (2) because suspended particulate matter is deposited in the lung and airways, it may interfere with breathing and hence affect performance; (3) since at the time of the study we did not know what effects oxidant had other than being an irritant, we had no clear-cut mechanism for thinking it may effect running.

Note that to help understand the possible relationships, we have arranged the results (that is the rows in the table) in descending order of percent of team with decreased performance. We now look at the upper five and lower five meets (Table 5) in terms of the percent with decreased performance, and we see that the mean values are shown in Table 5: (for some days, there were no data for the pollutants, so some of the averages are only for four numbers.)

We can see that the greatest proportional difference between the upper and lower five (or four) meets occurs with oxidant, the greatest absolute difference is for particulate matter and the least with carbon monoxide. We can use the scale of the SD as a way of judging the magnitude of the differences, since the mean and range of the pollutant levels vary among the three pollutants. For oxidant the difference is (26.5 to 6.4) 20.1, and SD for the

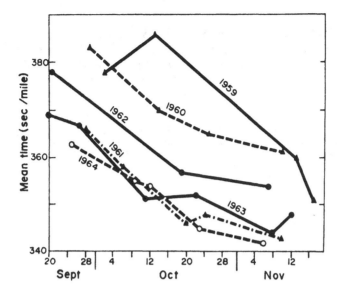

FIGURE 4. Mean running times of cross-country track teams during the seasons 1959 to 1964, San Marino, California. (From Wayne, W. S., Wehrle, P. F., and Carroll, R. E., *J. Am. Med. Assoc.*, 199, 901, 1967. With permission.)

whole set is 8.55. The difference is 2.35 times as great as the SD. For particulates the difference is 31.0 and the SD 18.94, for a ratio of 1.64. For carbon monoxide the difference is only 2.6, compared to a SD of 3.47 and the ratio is 0.75. *Such a finding is a basis for rejecting the hypothesis of an important effect of elevated carbon monoxide the hour prior to the race.*

Thus, we believe that the relationship of oxidant to performance is more impressive than that of particulate matter and that the influence of carbon monoxide, if any, is slight. We could also look at the extent that values above the median for performances are also associated with above median values for the pollutant levels an hour before the race.

However, the most impressive way to look at the relationship is graphically. We want to prepare what is called a scattergram, a plot of the percent of the team with decreased performance on the vertical axis and the pollutant level an hour before the race on the horizontal axis. When we do this we get the results shown in Figure 5. We can now see that the points nearly fall on a straight line for oxidant. (If they did all fall on a straight line the correlation coefficient, r, would be 1.00) The correlation coefficient for oxidant 1 hr before the race is in fact 0.868, for particulates 0.611, and for carbon monoxide 0.189. The correlation coefficients for oxidant 2 hr before the races and 3 hr before, were lower. This makes good sense. The investigators also looked at the average pollutant levels during the first half of the seasons and the second half and found them nearly identical, so the expected confounding by time of year was not likely to have affected these results. Neither did they find that daily averages of the pollution levels was associated with the performance, nor was day of the week. (Of course all of the races were on school days, so weekend to weekday differences would not be affecting the results.)

A question frequently asked is how little an amount of a pollutant is needed in order to produce an effect on health. With a correlation as strong as the one observed here, there may be an inclination to feel that the effect begins at a concentration as low as the lowest one which is graphed, in this case, 3 parts per hundred million, or 0.03 ppm. This would be a mistake. A simple, and very practical way to avoid such an error is to cover the right hand portion of a scattergram, and then one sees that, for example, if one looks at points

Table 4

**ANALYSIS MADE OF MEETS BASED ON PERCENT OF
THE TEAM WHOSE PERFORMANCE WAS POORER ON
A GIVEN DAY THAN ON THE MEET IMMEDIATELY
PRECEDING**

Percent of team with decreased performance	Oxidant 1 hr before race (pphm)[a]	Particulates 1 hr before race (km × 10)	Carbon monoxide 1 hr before race (ppm)
79	28	80	—
63	30	61	7
58	29	37	14
52	—	—	17
50	19	38	14
46	20	23	17
40	20	43	—
30	9	43	16
27	22	22	13
22	—	78	14(M)
19(M)	15(M)	41	14
19	7	33(M)	9
18	17	40	17
18	9	23	18
16	12	20	12
16	7	18	11
15	3	17	6
10	4	40	14
9	10	21	13
7	7	21	9
0	8	16	10

Mean	29.238	14.526	35.75	12.895
SD	21.204	8.553	18.943	3.478

[a] The units for oxidant are (pphm), parts per hundred million by volume, for particulates an arbitrary unit of transmittance km multiplied by 10, and for carbon monoxide parts per million, (ppm) (M) indicates the median, namely the value above which half of the values occur.

below 20 pphm there does not appear to be any relationship at all. It is only those points reflecting exposures above 20 pphm, together with the remainder, which begin to indicate a strong relationship.

One must also beware of overinterpreting scattergrams, which may have a high and statistically impressive correlation coefficient, but in which only one or two high exposure and effective relationships are represented, and the remainder appear to be nondescript. *In order for correlation analysis to provide a satisfactory basis for sound interpretations, there must be an approximately uniform distribution of the points on both axes.*

To summarize these numerical analyses, the most impressive results are the scattergrams, which "hit us between the eyes" with an almost linear relationship between oxidant levels the hour before the race and performance, in contrast to the lack of such a relationship with carbon monoxide, which we had thought would show an effect in advance. Particulate matter levels also seem to have some connection with performance.

This discussion is only a small example of a few principles for looking at numerical relationships between pollution and health of groups of people. The study of Biostatistics

Table 5

UPPER 5 AND LOWER 5 MEETS IN TERMS OF PERCENT WITH DECREASED PERFORMANCE

Mean pollution levels	Oxidant	Particulates	Carbon monoxide
For the 5 meets with highest proportion with decreased performance	26.5	54.0	13.9
For all meets	14.53	35.75	12.9
For the 5 meets with lowest proportion with decreased performance	6.4	23.0	10.4

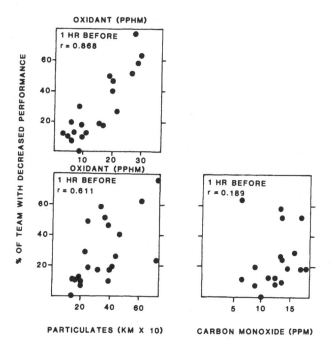

FIGURE 5. Scattergrams of proportions of the track teams times which show poorer performances compared to the last meet, compared to oxidant levels 1 hr prior to the meet (upper panel), particulate matter 1 hr prior to the meet (lower left panel), and carbon monoxide levels 1 hr prior to the meet (lower right panel). (From Wayne, W. S., Wehrle, P. F., and Carroll, R. E., *J. Am. Med. Assoc.*, 199, 901, 1967. With permission.)

and Epidemiology is intended to provide a comprehensive view of these relationships. (For references see Section X.)

Résumé of Section VII.

● In looking at a possible relationship between pollutant exposures and health reactions which may be a consequence of such exposures, we must consider and exclude possible sources of spurious association.

- We may want to transform the type of data from measurement to counting (or vice versa) in order to reduce the risk of finding spurious (that is noncausal) relationships. It is useful to array the data sets in order of increasing or decreasing values of an outcome variable.

- Graphical analysis is the most suitable for looking at possible relationships for their visual impact.

- Looking for the differences in exposures between the upper and lower groups of an ordered distribution, and examining the ratio between their mean differences and the SD of the whole set is a useful approximate method.

- Use of the median value and looking for the agreement between the values of the effect and exposure above and below the median is another approximate method.

- The correlation coefficient is a useful numerical criterion of whether or not there is a linear relationship within the entire range of observations, but does not necessarily reflect the dose-response relationship within any part of the range.

- The correlation coefficient, no matter how high, does not reflect the occurrence of beginning effects at the lowest level of pollution included in the scattergram.

- Scattergrams with the points clustered around low values and with one or two values reflecting high exposures and effect must be viewed with caution. Obscuring the right hand portion of a scattergram can help to avoid misinterpretations.

VIII. ETHICAL ISSUES AND MOTIVATION IN ENVIRONMENTAL EPIDEMIOLOGY

Several basic principles have been widely adopted for protecting human subjects in research; these principles have been developed to keep enthusiastic scientists constantly aware of the possibility that their studies may have harmful results for the subjects and to assure that the risk of such harm is kept as low as possible. From a superficial point of view it may seem that such principles are not relevant to asking people questions about their health and about the possible environmental exposures they may have experienced. While it is true that serious injury is less likely to occur as a result of looking for records in files and of interviewing persons than for instance using a new drug or operation, it nevertheless follows that similar principles affect epidemiological studies as affect other investigations that involve humans.

The first of these cardinal principles is that of "informed consent". It holds that persons whose involvement one wants must be told what the purposes of the study are and what possible risks they face if they participate. They are then asked to consent to participate and assured that if they wish to terminate their involvement at any time they may do so with no cost or no obligation to explain the reasons for their decision.

The second principle is that the risk of harm must be kept to a minimum and such risks must be justified by the potential benefit.

The third principle is that all information about a person's health is his or hers to keep private, and no disclosure of such information may be made without their consent.

These principles may lead to some conflicts in the conduct of environmental epidemiological studies, and a description of such possible conflicts and means for resolving them are suggested.

Two kinds of conflict may involve the principle of "informed consent". The first is that when we have a target population for study we want to get response information from as large a proportion of them as possible. If, for example, we have only response information for 60% of the target population, then we have no way of knowing if that 60% is respesentative or whether the other 40% has much more difficulty or much less. We usually try to get over 90% response rate, but the usual rate of response to a single mailing is about 50%, so we may then send further mailings and follow up the remaining non-respondents with phone calls or personal interviews. We may be tempted to use some form of pressure to increase the response rate, but to do so would violate the principle of "informed consent" which requires voluntary participation. The best way to resolve this conflict is to make participation in the study as attractive and as desirable as possible, by having support of key groups, introducing the study by letter or authority of a respected person in the community, and by clearly defining the possible benefits of participation both for the individual and for the community as a whole.

The second type of possible conflict with the principles of "informed consent" involves what is told to the respondents as to the purposes of the study. If we are, for example, telling our respondents that we are trying to find out whether exposure to pulp mill fumes is related to increased frequency and severity of sinus trouble, which may be one of the objectives of a study, then the respondents who usually try to give "helpful" answers may unconsciously bias the results of the study by recalling everything that could possibly "interest" us. Then, once the results are available and our colleagues review our methods, we can be criticized for asking leading questions or biasing our respondents. For these reasons, it is preferable to give general but not specific objectives of the study, for example introducing it as a study of environmental factors in respiratory disease, which of course obliges us to ask about a variety of respiratory disease and not to show that answers related to air pollution are more helpful or useful than answers about weather changes, smoking, fog, heat and cold spells, and wind direction. So to resolve this possible conflict, we always tell the respondents what our objectives are, but not in terms specific enough that they could give biased information.

The types of data we seek to collect in community environmental health studies does not usually involve much risk for the respondents, but some questions as to the occurrence of disease may provoke anxiety, and we should recognize such risks and do what we can to minimize them.

It is wisest to adopt the conservative position that no one should be asked to provide information who does not have a clear picture of what he and his community may gain from his (or her) cooperation. Each participant or potential participant should be considered as having a personal stake in the study and both the actions and attitude of the study staff should reflect this.

For many reasons, including costs, it is often desirable to use data already available, some of which are related to health or diseases which members of the community have had. Such information as people provide to their doctor concerning their health is for the purpose of helping their doctor prevent or treat disease, and is considered "privileged information", meaning that it may not be disclosed to other persons without the consent of the patient. A similar restraint applies to information concerning hospitalizations. Unless we obtain the permission of the patient, we cannot use information about his or her health experiences from other sources. Often, we really do not need to be able to relate the health data we want to analyze to an individual person, but we may only want to know the number and severity of cases of skin allergy, asthma, or some such possible indicator of health reaction. In this event, we should be willing to accept data which cannot be identified with a given person, but only, for example to a neighborhood. Then considerations of privacy and confidentiality are not a problem. Occasionally although we never really need to know the identity of each individual, it is important to use identifying information for purposes of record linkage, so one of the members of the study team does know the name and address of those affected. When this occurs, such identifiable information should be kept in a locked cabinet and access restricted only to a trained professional staff who can be counted on to maintain the records as confidential.

In a well planned and conducted study, respondents can be expected to understand and to cooperate in many helpful ways with the study staff, once they are convinced of the helpful intentions of the people who contact them. This can be assured by having the written support of the local health department, environmental health unit, medical society, and voluntary health associations. In the case of a study involving school children, the support of the parent-teachers group and the school board is also very important. Especially when one of the local industries is a suggested or obvious source of pollution, the field staff should avoid premature blame or criticism; some of the possible respondents may be employed at the plant in question, and have very complex feelings as to whether the operation is or is not causing health problems in the community. Interviewers or phone contacts who are good listeners are a big help; those who seem to have made up their minds as to a troublesome problem in the community *prior to doing a study of it* will have a hard time obtaining consistent cooperation. The investigator who has no feelings is not a very good investigator, but an investigator who shows his feelings at the wrong time is not likely to get much cooperation. The use of trained interviewers is one way to avoid the possibility that the investigator may let his or her feelings unfavorably affect either the cooperation of the subjects or the validity of the study findings.

Access to health information in the community is the most valuable asset of epidemiologists. It can be protected if each survey in the community is well planned, conducted at times and in circumstances convenient to the respondents, and if the persons who do such surveys have the courtesy of reporting back to the respondents, and showing that their cooperation is appreciated. A letter of appreciation to respondents is often a very prudent investment, particularly if some test has been done and the results can be of use or reassurance with respect to some health matter which was explored by the interview process.

In a community in which there are strong differences in opinion as to the environmental health hazards and what should be done about them, it is essential for the interview and field staff to maintain an open mind and a neutral position for two reasons. Cooperation of respondents must not be restricted by some of them feeling that they are talking to a "friendly" group and others feeling they are talking with the opposition. Secondly, the purposes of environmental health studies is to find information which the community may use to protect itself. Studies usually cannot be repeated, so it becomes of great importance that the first effort be undertaken with the greatest openmindedness and on the basis of thorough planning and preparation. Completeness of the input information and credibility of the study and its interpretation are at stake, and any lack of objectivity can jeopardize the whole effort.

Institutional backing and support can make the study easy or difficult, and considerable effort is justified in order to have the support the study deserves; often this means waiting for some sort of administrative review. Among the most important types of such review are the so-called "Human Studies Committee" review procedures which are intended to assure that the principles for protection of human subjects in research are complied with. Although the extra paper work and possible delay in such reviews may be burdensome, it is better to go patiently through the procedures which a given institution may require rather than to run the risk, that, due to impatience, an avoidable problem becomes a serious obstacle to the conduct of the research.

Résumé of Section VIII.

- Three basic ethical principles should guide community studies of environmental health problems as they do all other research involving human subjects: the principle of informed consent, the principle of avoiding risk of harm to the participants, and the confidentiality of medical records.
- Conflicts which may arise with respect to these principles can and must be resolved. The principle of informed consent may seem to limit the chance to get a high level

of participation, but thoughtful and effective planning can maintain high participation without compromising the voluntary nature of participation.

- While subjects must be informed as to the objectives of the study, they usually need not be told them in ways which could bias their responses.

- Since the personal identity of respondents is usually of no special value for such studies, it is preferable to avoid using records which contain personal identifying information.

- Institutional review procedures, though appearing tedious, usually justify themselves when it is realized how valuable institutional support may be.

- Two principles are strongly recommended as a basis for assuring community participation; first objectivity and neutrality concerning any related matter about which there are differences of opinion in the community; secondly, the study participants should have and should be treated as though they have a personal stake in the success of the study.

IX. PLANNING STRATEGIES

There are three phases in the planning of any community study. The first phase is obtaining of the consent or permission of key persons or organizations; sometimes this is done in writing, and sometimes in person or by phone. The second phase is the preparation of a research proposal which under most circumstances is in the form of a request for funds or their equivalent. The third phase begins when the likelihood of actually doing the study becomes real; it involves the preparation of what is called a protocol for the study. The fourth phase is the execution of the work called for in the research proposal along the lines set down in the protocol. Throughout these phases and the work and preparation they imply, run the themes and questions involved in the aims and methods of the research.

Let us take a hypothetical problem and follow its hypothetical course through these phases. After an unusually cold winter preceding a general election, it has been said in a newspaper story that during this winter there were more cases of pneumonia and serious respiratory disease, and that the infant mortality rate had increased in a major city, partly as a result of energy conservation efforts and high unemployment. We have decided to undertake the study, knowing that in some buildings, landlords have failed to provide sufficient heat for their tenants, and notwithstanding the possible political motivation of the newspaper story, the research might make a valuable contribution to knowledge about housing conditions and health.

Our objectives are (1) to see whether there were more cases of pneumonia among persons with inadequately heated housing this winter than in previous winters; (2) to see if infant mortality was greater this winter amoung families in inadequately heated housing than it was during previous winters; (3) to find out waht the relative magnitude of serious respiratory illness was among persons with inadequately heated housing this winter as compared to previous winters.

In order to acheive these objectives, we must: define a population which lived in inadequately heated housing this winter for whom data for experiences during previous winters is also available and for whom an "otherwise comparable population which has had no inadequacy of heat this winter is available for comparison". We must have access to infant mortality data by place of residence and time of year for this as well as previous winters. We must be able to interview a sample of people both as to the adequacy of heat during the winter period and as to the occurence of respiratory illness during this as well as previous winters. Since recall bias is likely to make recall for the immediate past period better than for previous periods, we shall need to have available some source of records as to the incidence of pneumonia and serious respiratory illness for the population groups we intend to study.

Now for the first phase of planning, we will need to have access to census data in order to draw a sample, or we will need to take our own census, a relatively costly procedure we would rather not have to undertake. Let us assume that we have access to a recent set of census records from the library (no permission required). We would like to examine records of fuel or power suppliers in order to try to determine if fuel or power were discontinued in any of the buildings in our study area. We shall want recorded data from the Weather Bureau as to temperature ranges, wind, and precipitation as well as "degree days" data for several previous years (also probably available at the library).

We shall want data on births and infant mortality, which probably can be provided by the local health department, but we shall be willing to assure them that the data can be useful to us without any personal identification if that will make it easier to provide; we shall want the address, or course. (This may take some negotiations and possibly a fee for copying the data.) We will want to obtain data on cases of pneumonia seen at local hospitals and emergency medical services for several years, and again, as long as we have the address, we do not need to have any personal identifying information. (This too, will require some negotiations and may require payment to record librarians.) Finally, we will want to be able to interview the sample of persons in order to determine the occurrence of "severe respiratory illness" and how much and when they were exposed to cold as a result of inadequate heating. Possibly there is a local community council, a tenant's association, whose cooperation may be vital to our access, but it may also want to be sure that it recruits and we hire interviewers from the community. We may need the approval of the local city council, the local health department, and the local medical association. *The essential point is that we need to know whom to address and how to address the individuals and organizations whose cooperation we need.* One of our objectives is their support and agreement with the validity of our proposal and the qualifications we have to undertake it. As second objective is to get information as to how much money we shall need in order to conduct the study.

We assume that we have received the critical backing we have requested as well as a possibly inflated set of requests for money. We now face the task of requesting the resources we shall need to carry out the study. This is a classical "grant application" procedure, which includes a background statement, a short list of "aims" or "objectives of the research", a section on methods of achieving the aims, a budget, a statement as what previous work has been done on this and related matters, possibly a list of references, and a statement as to the significance of the study. A "Schedule of the Work to be Done" is an important part of the proposal. This grant application should note the problems expected to be encountered and the methods which you propose for overcoming them, but this is for the purpose of assuring the granting authority that you are realistic and competent.

When submitting such a proposal, you should have in mind what is the realistic minimum of support you absolutely must have if any useful information is to be obtained, because often the amounts requested are not granted. If this is the case, you may decide that since the infant mortality data will probably not cost you much, that the minimal project will involve only the single objective and that you will then keep searching for enough funds to do a more comprehensive job. Possibly, you may feel that you can interest one of the doctors in the chest clinic at the local hospital to do the work on the occurence of pneumonia, and therefore by collaborating with him, you may not have to support that part of the data collection from funds you obtain. This phase of the planning, of course, is confidential; it nevertheless should not be omitted!

The final document is essentially the guide to how the study should be carried out. It contains copies of all questionnaires, dummy tables of the data you expect to get, statistical and logical criteria for interpretation, draft letters of introduction and appreciation. It also should contain in as much detail as possible the problems you expect to encounter and the measures you believe will be useful for dealing with them. This working protocol is intended to be a looseleaf notebook, with job descriptions, schedules, staff and supply schemes, names addresses, and phone numbers of all relevant staff and consultants.

It is not intended to impress anyone, so you should be frank and complete as to the possible problems and strategies you may need.

An important part of the planning of the final study is the pretesting of all procedures you intend to carry out in the community. Each questionnaire and procedure must be tried out on people similar to those you expect to be dealing with.

Disseminating the results of the study to participants, to other investigators, and to those who may need to be involved in correcting any problems uncovered should be planned as part of the study plan itself. Too often, this is not included in the planning, and as a consequence, the impact of all of the work and participation is less than it should be.

The résumé of this section in effect is reflected in the exemplary studies presented in Section II.

X. SOURCES OF ADDITIONAL HELP

Community environmental health problems are reported in a variety of scientific journals and newsletters, which usually can be found in your local library. Textbooks on the basic scientific disciplines will be found useful. In addition, new books are beginning to appear which provide additional perspectives on the rapidly developing field of environmental epidemiology.

A. Periodicals

The following periodicals are used primarily to keep up to date, but can, if read systematically, provide good background information:

Science, the weekly publication of the American Association for the Advancement of Science, in its news and reports section provides excellent coverage of political developments, with occasional articles on such subjects as indoor air pollution, environmental exposures and cancer, and new methods for measurement.

Archives of Environmental Health, a monthly publication of Heldref Foundation, is the senior and most generally useful of the environmental journals for those concerned with community studies.

Environmental Research published by Academic Press is a monthly which includes both community, occupational health, and laboratory studies.

The American Journal of Public Health, a monthly publication of the American Public Health Association has a minority of articles on community environmental health, but has other articles on community health and action.

The American Journal of Epidemiology, a monthly, has relatively technical articles on all aspects of epidemiology, a small portion of which are on community environmental health issues.

The Environmental Health Reporter, published by Gerson Fishbein is a weekly letter-format publication which specialized in news of political processes as well as governmental budgets, contracts, and grants.

Environmental Health Perspectives, published by the National Institute of Environmental Health Sciences of the U.S. Department of Health and Human Services, published semimonthly, tends to publish in-depth reports or meeting conferences on scientific subjects.

Environmental Science and Technology is primarily concerned with engineering, monitoring, and control technology.

The American Review of Respiratory Diseases, published monthly by the American Thoracic Society, the medical arm of the American Lung Association, is primarily concerned with air pollution and its effects on diseases and conditions of the lung. It has kept a strong emphasis on community studies of such problems, partly as a result of the influence of the American Lung Association (the Christmas Seal Organization) which has a staff of trained community workers.

In other countries, the *Scandinavian Journal of Work, Environment and Health*, (published in Finland), *Science of the Total Environment* (published in The Netherlands), and the *Journal of Epidemiology and Community Health* (Great Britain) are major English language publications.

Environment in the U.S. and *Ambio* in Sweden are publications oriented to layman concerned with environmental quailty.

B. Monographs and Textbooks

Health Hazards of the Human Environment a multi-authored publication of the World Health Organization remains the single best one-volume introduction to environmental health problems, even though it was published in 1972.

The World Health Organization also published a number of Technical Reports, and a Bulletin from Geneva Headquarters, and increasingly from its Regional Offices, especially from the Copenhagen headquarters of its European Region.

A special issue of *Science of the Total Environment* on "Environmental Monitoring of Environmental Health Hazards" edited by J. R. Goldsmith was published in January 1984. It emphasizes the uses of epidemiology to assure that health problems in new technical developments are recognized and that prompt action is taken to abate them.

A general textbook entitled *Environmental Health by* P.W. Purdom, was published by Academic Press in 1980.

Academic Press also publishes *Air Pollution* a multi-volume, multi-authored set of books, edited by Arthur Stern.

Air Conservation represents the results of a Commission appointed by the American Association for the Advancement of Science, and published by the Association in 1965. It remains a model of how to deal with pollution from a conservationist point of view.

CRC Press published a book entitled *Occupational Epidemiology* by R. Monson, and a series of technically oriented reviews.

Textbooks of epidemiology include: *Uses of Epidemiology* by Morris, a classic now in its third edition, Churchill Livingstone (Longmans, Edinburgh, 1975). *Foundations of Epidemiology* by A. M. Lilienfeld and D. E. Lilienfeld, 2nd ed., Oxford, New York, 1980.

Epidemiologic Methods by MacMahon, Pugh, and Ipsen, Little, Brown and Co., Boston. *Principles of Epidemiology: a Self Teaching Guide* by L. H. Roht, B. J. Selwyn, A. H. Holguin, and B. L. Christensen, Academic Press, New York, 1982 responds to the need for a broader understanding of epidemiology.

Books dealing with the general subject matter of environmental epidemiology but with different perspectives include: *Environmental Epidemiology*, P.E. Leaverton, L. Masse, S. O. Simches, and D. M. Righi, Eds., Praeger, New York, 1982 is a collection of acticles rather than a systematic treatment.

"Statistics Needed for Determining the Effects of the Environment on Health", a report of a Technical Panel of the U.S. Committee on Vital and Health Statistics, U.S. National Center for Health Statistics, Department of Health and Human Resources, Vital Statistics Reports Series 4, #20 PHS HRA 77-1457, provides a background for the various sources of health statistical data in the U.S. relevant to environmental health.

Public Health and Preventive Medicine (Maxcy-Rosenau), 11th ed., J. Last, Ed., is a comprehensive multi-authored textbook of public health with excellent chapters on environmental and community health. It is published by Appleton-Century Crofts, New York, 1980.

The World Health Organization in Geneva has been at work for at least 5 years on a monograph on "Guidelines on Studies in Environmental Epidemiology" published as Environmental Health Criteria #27, Geneva, 1983. Its focus is on technical rather than community aspects of environmental health studies.

A new textbook on *Environmental Epidemiology: Principles and Practice* by G. H. Spivey and A. H. Coulson is being published by Ann Arbor Science Publishers (Butterworth Group). The book uses five case studies as a focus for discussion of methods and strategies. They are Subclinical Neurological Effects of Lead, Vinyl Chloride Exposures in Schoolchildren, "Agent Orange" (Dioxin) Exposures to Vietnam Veterans, Epidemiological Monitoring in the Vicinity of a Toxic Waste Site, and An Apparent Increase in Congenital Malformations Among Communities in which Brush Spraying had Occurred. This approach, a longitudinal one is complementary to the cross-sectional one used in the present book.

REFERENCES

1. **Hippocrates,** "Of Airs, Waters and Places" in *The Genuine Works of Hippocrates: Translated from the Greek,* Krieger, Huntington, N.Y., 1972, 19.
2. **Goldsmith, J. R. and Breslow, L.,** Epidemiological aspects of air pollution, *J. Air Pollut. Cont. Assoc.,* 9, 129, 1959.
3. **Waller, R. E.,** Control of air pollution: present success and future prospects, in *Recent Advanced in Community Medicine,* Bennett, A. E., Ed., Churchill Livingstone, Edinburgh, 1978, 59.
4. **Lawther, P. J., Waller, R. E., and Henderson, M.,** *Thorax,* 25, 525, 1970.
5. Health Statistics Needed for Evaluation of the Impact of Environmental Pollution on Health, a report of a panel of the U.S. National Committee on Vital and Health Statistics, Health Statistics Report Series #4, No. 20, U.S. Government Printing Office, Washington, D.C., PHS HRA 77-1457.
6. **Wayne, W. S., Wehrle, P. F., and Carroll, R. E.,** *J. Am. Med. Assoc.,* 199, 901, 1967; see also Section 2.10, p. 61.

Section II: Exemplary Studies
Part A

Chapter 2*

THE BROAD STREET PUMP

Inge F. Goldstein and Martin Goldstein

TABLE OF CONTENTS

*   Adapted from a chapter in *The Experience of Science: An Interdisciplinary Approach*, Martin Goldstein and Inge F. Goldstein, Plenum Press, New York, 1984.

I. INTRODUCTION

At the end of August 1854, in the neighborhood of Broad Street in the Soho district of London, there was a devastating outbreak of cholera. In 10 days, there were over 500 deaths among people who lived or worked within a 250-yard radius of the intersection of Broad Street and Cambridge Street. On the evening of September 7, the vestrymen of the St. James Parish, who constituted a local governing body for this district, were holding a meeting to consider what could be done about the outbreak and the resulting panic and exodus of the inhabitants. A stranger asked for permission to speak, and when given the floor, argued that the source of the disease was a pump located at the corner of Broad and Cambridge Streets that supplied the water for the affected neighborhood. He recommended that the handle of the pump be removed, so that the people living there would be forced to draw their water from other pumps located several streets away.

The vestrymen were somewhat skeptical, but agreed to do so. The pump handle was removed. This act was one of the most famous public health measures ever taken, and it accomplished very little. The epidemic had already passed its peak, and the removal of the pump handle seemed to have no clear-cut effect.

The stranger at the meeting was a London physician, John Snow. Snow had been studying cholera carefully since a major epidemic in London in 1849. He had come to the conclusion that cholera can be transmitted through the water supply if the water is contaminated with sewage bearing the excretions of cholera victims, and his study of the distribution of the cholera cases in this district and surrounding districts of London had led him to the conclusion that there was a leakage of sewage into the water supply of this particular pump.

How did Snow come upon his theory, what kind of evidence did he offer for it, what kinds of public health measures did he recommend, and what effect did this have on the occurrence of cholera and on public health generally?

II. THE STATE OF MEDICAL KNOWLEDGE

The concept of communicable diseases — that some diseases are transmitted by close contact from the sick to the well — came into being in the Middle Ages.[1] The ancient Greeks were the first to attempt to look at disease scientifically. They rejected the idea of disease as a punishment for sin or as a consequence of witchcraft, and studied instead the relation of diseases to aspects of the natural environment or the way men live, eat, and work. They noted, for example, that it was unhealthy to live near swamps. But in spite of the fact that they suffered from epidemics of various sorts, they somehow missed recognizing that some diseases are contagious.

The prescriptions for isolation and purification described in the Hebrew Bible for physiological processes such as menstruation and for diseases characterized by discharges or skin lesions apparently are based on the idea of the contagiousness of spiritual uncleanliness, of which the physical disease was merely an external symptom. In the Middle Ages, the Church, confronted with a major epidemic of leprosy, revived the biblical practice of isolation of the sick, and the same methods were applied during the outbreak of the Black Death (bubonic plague) in the 14th century. By this time the concept of contagion was well established.

It is interesting that the belief that disease was a consequence of evil behavior coexisted with the recognition of contagion for hundreds of years. Attempts to develop treatments for syphilis were opposed on the grounds that syphilis was a just penalty for sexual immorality. Cholera was most prevalent among the poor for reasons that will become apparent, and there were many who regarded it as a proper punishment for the undeserving and vicious classes of society. A governor of New York State once stated during a cholera epidemic, "...an infinitely wise and just God has seen fit to employ pestilence as one means of scourging the human race for their sins, and it seems to be an appropriate one for the sins of uncleanliness and intemperance..." The president of New York's Special Medical Council stated at the onset of an epidemic in 1832, "The disease had been confined to the intemperate and the dissolute with but few exceptions." A newspaper report noted, "Every day's experience gives us increased assurance of the safety of the temperate and prudent, who are in circumstances of comfort... The disease is now, more than before, rioting in the haunts of infamy and pollution. A prostitute at 62 Mott Street, who was decking herself before the glass at 1 o'clock yesterday, was carried away in a hearse at half past three o'clock. The broken down constitutions of these miserable creatures perish almost instantly on the attack... But the business part of our population, in general, appear to be in perfect health and security". A Sunday school newspaper for children explained: "Drunkards and filthy wicked people of all descriptions are swept away in heaps, as if the Holy God could no longer bear their wickedness, just as we sweep away a mass of filth when it has become so corrupt that we cannot bear it... The Cholera is not caused by intemperance and filth in themselves, but is a scourge, a rod in the hand of God".[2]*

Today we know that cholera is a bacterial disease. It is characterized by severe diarrhea, vomiting, and muscular cramps. The diarrhea can produce extreme dehydration and collapse; death is frequent, and often occurs within hours after the onset of sickness.

The disease had been known to exist in India since the 18th century, and occurs there and in other parts of the world today. In the 19th century, as travel between Asia and the West became more common and as the crowding of people in urban centers increased as a result of the industrial revolution, major epidemics occurred in Europe and America. England had epidemics in 1831-32, 1848-49, and 1853-54.

The question of how cholera is transmitted was especially difficult. On the one hand there was good evidence that it could be transmitted by close personal contact. Yet there was equally good evidence that some who had close personal contact with the sick, such as physicians, rarely got it, and that outbreaks could occur at places located at great distances from already existing cases of the disease.

A number of theories were proposed, some of which were too vague to be rationally examined but some of which had solid experimental support.

The transmission of contagious diseases in general and cholera in particular was frequently explained by "effluvia" given off in the exhalations of the patient or from bodies of the dead, and subsequently inhaled into the lungs of a healthy person.

It is easy today to look down on the effluvia theory as so much unenlightened superstition. One should recognize, however, that the germ theory was then highly speculative and had very little evidence in its favor. The idea that disease could be spread by foul odors or other poisonous emanations represented a great advance over views attributing disease to witchcraft or sin, and, in the absence of any knowledge of microorganisms, was a plausible explanation of contagion.

Further, the effluvia theory led to justified concern over the crowded and unsanitary living and working conditions of the poor. Interested readers should consult the report prepared for Parliament by E. Chadwick in 1842 for a description of these conditions.[3] Chadwick's report led to the first serious public health measures taken by the British government, and in fact these measures resulted in improved health of the population of England.

This illustrates a truism of scientific research: an incorrect theory is better than no theory at all, or, in the words of an English logician Augustus de Morgan, "Wrong hypotheses, rightly worked, have produced more useful results than unguided observation".[4]

---

* These quotations are selected from many given by Charles E. Rosenberg in *The Cholera Years*, University of Chicago Press, Chicago, Ill., 1962.

A number of people, including both physicians and uneducated laymen, had blamed the water supply. Snow adopted this theory, but refined it by specifically implicating the excretions of the cholera victims.

He noted that the very first symptom of the disease is a severe and uncontrollable diarrhea that leads to collapse and often death. There are no preliminary signs of fever or sensation of illness before this. He reasons as follows:

As cholera commences with an affection of the alimentary canal, and as we have seen that the blood is not under the influence of any poison in the early stages of this disease, it follows that the morbid material producing cholera must be introduced into the alimentary canal — must, in fact, be swallowed accidentally, for persons would not take it intentionally; and the increase of the morbid material or cholera poison must take place in the interior of the stomach and bowels. It would seem that the cholera poison, when reproduced in sufficient quantity, acts as an irritant on the surface of the stomach and intestines, or, what is still more probable, it withdraws fluid from the blood circulating in the capillaries, by a power analogous to that by which the epithelial cells of the various organs abstract the different secretions in the healthy body. For the morbid matter of cholera having the property of reproducing its own kind, must necessarily have some sort of structure, most likely that of a cell. It is no objection to this view that the structure of the cholera poison cannot be recognized by the microscope, for the matter of small-pox and of chancre can only be recognized by their effects, and not by their physical properties. (pp. 16—17)*

Having guessed that the excretions of the victims are the means by which the disease is transmitted, he makes a number of sociological observations that explain why the disease spreads more rapidly among the poor than among the rich.

Nothing has been found to favour the extension of cholera more than want of personal cleanliness, whether arising from habit or scarcity of water, although the circumstances till lately remained unexplained. The bed linen nearly always becomes wetted by the cholera evacuations, and as these are devoid of the usual colour and odour, the hands of persons waiting on the patient become soiled without their knowing it; and unless these persons are scrupulously cleanly in their habits, and wash their hands before taking food, they must accidently swallow some of the excretion, and leave some on the food they handle or prepare, which has to be eaten by the rest of the family, who, amongst the working classes, often have to take their meals in the sick room: hence the thousands of instances in which, amongst their class of the population, a case of cholera in one member of the family is followed by other cases; whilst medical men and others, who merely visit the patients, generally escape. (pp. 16—17)

The relative rarity of cholera among physicians attending victims of the disease is noted by Snow as evidence against the "effluvia" theory:

The low rate of mortality amongst medical men and undertakers is worthy of notice. If cholera were propagated by effluvia given off from the patient, or the dead body, as used to be the opinion of those who believed in its communicability; or, if it depended on effluvia lurking about what are by others called infected localities, in either case medical men and undertakers would be peculiarly liable to the disease; but, according to the principles explained in this treatise, there is no reason why these callings should particularly expose persons to the malady. (p. 122)

Further evidence against the effluvia theory is the fact that cholera sometimes breaks out during an epidemic in new areas remote from other cases of the disease. It is this observation that Snow could plausibly explain as the result of contamination of the water supply by sewage. He does not claim credit for discovering this hypothesis, but quotes anecdotal observations by others:

Dr. Thomas King Chambers informed me, that at Ilford, in Essex, in the summer of 1849, the cholera prevailed very severely in a row of houses a little way from the main part of the town. It had visited every house in the row

---

* Quotations from Snow have been taken from Snow, J., *Snow on Cholera*, The Commonwealth Fund, New York, 1936. (Page references are to this edition.)

but one. The refuse which overflowed from the privies and a pigsty could be seen running into the well over the surface of the ground, and the water was very fetid; yet it was used by the people in all the houses except that which had escaped cholera. That house was inhabited by a woman who took linen to wash, and she, finding that the water gave the linen an offensive smell, paid a person to fetch water for her from the pump in the town, and this water she used for culinary purposes, as well as for washing.

The following circumstance was related to me, at the time it occurred, by a gentleman well acquainted with all the particulars. The drainage from the cesspools found its way into the well attached to some houses at Locksbrook, near Bath, and the cholera making its appearance there in the autumn of 1849, became very fatal. The people complained of the water to the gentleman belonging to the property, who lived at Weston, in Bath, and he sent a surveyor, who reported that nothing was the matter. The tenants still complaining, the owner went himself, and on looking at the water and smelling it, he said that he could perceive nothing the matter with it. He was asked if he would taste it, and he drank a glass of it. This occurred on a Wednesday; he went home, was taken ill with the cholera, and died on the Saturday following, there being no cholera in his own neighborhood at the time...(pp. 31—32)

III. THE BROAD STREET PUMP

Snow made a number of observations during the 1849 epidemic of an anecdotal type himself, that gave support to his hypothesis. By the time of the 1853-54 epidemic he was ready to subject it to a more exacting test. Here is his own description of the Broad Street-Golden Square epidemic:

The most terrible outbreak of cholera which ever occurred in this kingdom is probably that which took place in Broad Street, Golden Square, and the adjoining streets, a few weeks ago. Within 250 yards of the spot where Cambridge Street joins Broad Street, there were upwards of 500 fatal attacks of cholera in 10 days. The mortality in this limited area probably equals any that was ever caused in this country, even by the plague; and it was much more sudden, as the greater number of cases terminated in a few hours. The mortality would undoubtedly have been much greater had it not been for the flight of the population. Persons in furnished lodgings left first, then other lodgers went away, leaving the furniture to be sent for when they could meet with a place to put it. Many houses were closed altogether, owing to the death of the proprietors; and, in a great number of instances, the tradesmen who remained had sent away their families so that in less than six days from the commencement of the outbreak, the most afflicted streets were deserted by more than three-quarters of their inhabitants.

There were a few cases of cholera in the neighbourhood of Broad Street, Golden Square, in the latter part of August; and the so-called outbreak which commenced in the night between the 31st of August and the 1st of September, was, as in all similar instances, only a violent increase of the malady. As soon as I became acquainted with the situation and extent of this irruption of cholera, I suspected some contamination of the water of the much-frequented street-pump in Broad Street, near the end of Cambridge Street; but on examining the water, on the evening of the 3rd September, I found so little impurity in it of an organic nature, that I hesitated to come to a conclusion. Further inquiry, however, showed me that there was no other circumstance or agent common to the circumscribed locality in which this sudden increase of cholera occurred, and not extending beyond it, except the water of the above mentioned pump...(pp. 38—39)

Snow began his study by obtaining from the London General Register Office a list of the deaths from cholera in the area occurring each day. These figures showed a dramatic increase in cases on August 31st, which he therefore identified as the starting date of the outbreak. He found 83 deaths that took place from August 31 to September 1 (see Table 1), and made a personal investigation of these cases.

On proceeding to the spot, I found that nearly all the deaths had taken place within a short distance of the pump. There were only ten deaths in houses situated decidedly nearer to another street pump. In five of these cases the families of the deceased persons informed me that they always sent to the pump in Broad Street, as they preferred the water to that of the pump which was nearer. In three other cases, the deceased were children known to drink the water; and the parents of the third think it probable that it did so. The other two deaths, beyond the district which this pump supplies, represent only the amount of mortality from cholera that was occurring before the irruption took place.

With regard to the deaths occurring in the locality belonging to the pump, there were 61 instances in which I was informed that the deceased persons used to drink the pumpwater from Broad Street, either constantly or occasionally. In six instances I could get no information, owing to the death or departure of everyone connected with the deceased individuals; and in six cases was informed that the deceased persons did not drink the pump-water before their illness...(pp. 39—40)

Table 1
RESULTS OF SNOW'S INVESTIGATION

83 Deaths[a]

73 Living near Broad St.			10 Not living near pump		
61	6	6	5	3	2
Known to have drunk pump water	Believed not to have drunk pump water	No information	In families sending to Broad St. pump for water	Children attending school near pump	No information

[a]   Out of 83 individuals who had died of the disease, 69 were known definitely or could be assumed to have drunk the pump water, 6 were believed to have drunk it, and for 8 there was no information.

A. Who Drank the Pump Water?

For reasons of clarity we summarize the results of Snow's investigation of these 83 deaths in Table 1, which shows that there were deaths among people not known to have drunk water from the Broad Street pump. These deaths therefore are facts that seem to contradict Snow's hypothesis. A scientist faced with facts contradictory to a hypothesis has many alternatives, only one of which is to discard the hypothesis. Another alternative is to check the "facts" to see if they really are the facts in this case. Still another is to make a closer examination of them, to see whether in some plausible way they can be shown either not really to contradict the hypothesis or actually to support it. It occurred to Snow to look for ways the individuals in question might have drunk the water without being aware of it.

The additional facts that I have been able to ascertain are in accordance with those above related; and as regards the small number of those attacked, who were believed not to have drunk the water from Broad Street pump, it must be obvious that there are various ways in which the deceased persons may have taken it without knowledge of their friends. The water was used for mixing with spirits in all the public houses around. It was used likewise at dining rooms and coffee shops. The keeper of a coffee shop in the neighbourhood, which was frequented by mechanics, and where the pump-water was supplied at dinner time, informed me (on 6th September) that she was already aware of nine of her customers who were dead. The pump-water was also sold in various little shops, with a teaspoonful of effervescing powder in it, under the name of sherbet; and it may have been distributed in various other ways with which I am unacquainted. The pump was frequented much more than is usual, even for a London pump in a populous neighbourhood. (pp. 41—42)

Snow gives two striking additional observations that confirm the role of the pump. There were two large groups of people living near the Broad Street pump who had very few cases of cholera: the inhabitants of a workhouse and the employees of a brewery.

In the workhouse, which had its own water supply, only 5 out of 535 inmates died. If the death rate had been the same as in the surrounding neighborhood, over 100 would have died.

He questioned the proprietor of the brewery, who informed him that there were no clear-cut cases of cholera among the 70 employees, who while on the job were allowed to drink a certain amount of malt liquor. To the proprietor's knowledge, none drank the pump water.

B. The Pump Handle

On September 8, the handle of the pump was removed, but, as Snow notes, by this time the epidemic had subsided, perhaps because many inhabitants had fled the neighborhood. So the removal of the pump handle did not produce any dramatic effect on the number of new cases (Figure 1).

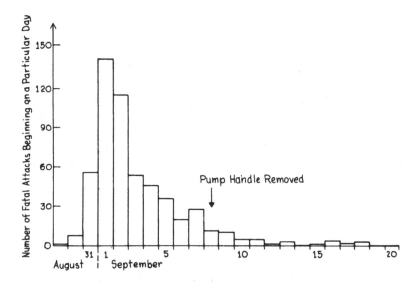

FIGURE 1.   The Broad Street Pump outbreak. The figure shows the number of fatal cases that began on a given date, plotted against the date. The arrow indicates when the pump handle was removed.

Following the epidemic, the pump was opened and examined. No direct evidence of leakage from nearby privies was found, but Snow states his belief that it must have occurred, perhaps by seepage through the soil, as on microscopic examination "oval animalcules" were found, which Snow points out are evidence of organic contamination. (They were not the bacteria causing cholera, which were not detectable by the microscopic techniques of the time, nor did Snow take them seriously as a causative agent — rather, he knew that "animalcules" were very common in natural waters contaminated with sewage or other organic matter, even when no cholera was present.)

Additional evidence for the contamination of the pump water with sewage was provided by inhabitants of the neighborhood who had noticed a disagreeable taste in the water just prior to the outbreak and a tendency of the water to form a scum on the surface when it was left to stand a few days. Further, chemical tests showed the presence of large amounts of chlorides, consistent with contamination by sewage, but, like the animalcules, not constituting overwhelming proof. The question of chlorides in the drinking water will come up again more dramatically later on.

Snow's conclusion on the Broad Street pump outbreak is as follows:

Whilst the presumed contamination of the water of the Broad Street pump with the evacuations of cholera patients affords an exact explanation of the outbreak of cholera in St. James parish, there is no other circumstance which offers any explanation at all, whatever hypothesis of the nature and cause of the malady be adopted...(pp. 51—54)

IV. INTERVENTION

The removal of the pump-handle represented what is called in modern public-health parlance an "intervention". Had it had a significant effect on the course of the epidemic, it would have provided very strong confirmatory evidence for the hypothesis that provided its rationale.

Examples of intervention that have been more successful, and thus more convincing, are well known: they include the cessation of cigarette smoking, which is followed by a greatly diminished risk of lung cancer, and measures to cut down industrial exposure to naphthylamine and other agents implicated in bladder cancer, after the adoption of which bladder cancer among workers in chemical industries has greatly diminished.

V. A CONTROLLED EXPERIMENT

Snow's analysis of the Broad Street-Golden Square outbreak in terms of his hypothesis is plausible and convincing, especially to us today, because we know he was basically correct. But the evidence is ancedotal: his hypothesis is consistent with most of the evidence he was able to obtain by observing the circumstances, and no alternative hypothesis appears to account for these observations as well, but a controlled experiment would be more convincing.

We are, of course, debarred by ethical considerations from deliberately exposing human beings to agents which might cause disease, and laboratory animals are not known to suffer from cholera.

The ethical problems can be avoided if by chance a "natural" experiment is available: it may happen that a group in the population has been exposed fortuitously to what is believed to be the cause of a disease. A controlled experiment is then possible if another group in the population can be found, similar in every relevant respect to the first one, except that it has not been exposed to the suspected cause. If the disease occurs in the first group and not the second, we have confirming evidence that the suspected cause really is the cause. The search for such "natural" experiments is the basic strategy of epidemiology. In such a "natural" situation it may be hard to prove the two groups similar in "every relevant respect". For example, different districts of London had different water supplies and different cholera rates. But unfortunately, from the point of view of testing Snow's hypothesis that cholera is caused by contaminated water, the people in the different districts were different in other ways, also. The rich lived in different neighborhoods from the poor and suffered less from cholera. Was it because they had uncontaminated water supplies or because they ate better food, worked shorter hours at easier jobs, lived in newer, cleaner houses?

Also, different groups of equally "poor" people might differ in other significant ways. In London at that time there was a tendency for people of the same occupation to live in a single neighborhood, so that one neighborhood might have a lot of butchers, another might have tailors, and a third drivers of carts. Might susceptibility to cholera depend on occupation? Snow himself was aware that some occupational groups such as doctors were less likely to get cholera, and some, such as coal miners, were more likely. Perhaps some overlooked causative factor was related to one's work.

Since we now know that Snow's theory about the water supply was correct, we may feel that all these other differences are irrelevant and can be disregarded. But at the time this wasn't yet clear, and of course, the purpose of the experiment was to find this out. If the control and test groups differed in three or four other ways besides getting their water supplies from a different source, we would not feel safe in blaming the water supply alone; any of these other differences between the groups might be responsible for the differences in cholera rates.

It was Snow's genius to recognize the importance of the fortuitous circumstance that two different water companies supplied a single neighborhood in an intermingled way.

The two water companies in question both drew their water from the Thames, originally from spots that could be expected to be contaminated with the sewage of the city. But in 1852, after the epidemic of 1849, one of these companies, the Lambeth Company, moved its waterworks upstream to a place free of London sewage. The other, the Southwark and Vauxhall Company, remained where it was. Both companies delivered drinking water to a single district of the city:

The pipes of each Company go down all the streets, and into nearly all the courts and alleys. A few houses are supplied by one company and a few by the other, according to the decision of the owner or occupier at that time when the Water Companies were in active competition. In many cases a single house has a supply different from that on either side. Each Company supplies both rich and poor, both large houses and small; there is no difference either in the condition or the persons receiving the water of the different companies.

In the next sentence, Snow summaries the basic idea of the experiment:

As there is no difference whatever, either in the houses or the people receiving the supply of the two Water Companies, or in any of the physical conditions with which they are surrounded, it is obvious that no experiment could have been devised which would more thoroughly test the effect of water supply on the progress of cholera than this, which circumstances placed ready-made before the observer.

The experiment, too, was on the grandest scale. No fewer than 300,000 people of both sexes, of every age and occupation, and of every rank and station, from gentlefolks down to the very poor, were divided into two groups without their choice, and, in most cases, without their knowledge, one group being supplied with water containing the sewage of London, and, amongst it, whatever might have come from the cholera patients, the other group having water quite free from such impurity.

To turn this grand experiment to account, all that was required was to learn the supply of water to each individual house where a fatal attack of cholera might occur. I regret that, in the short days at the latter part of last year, I could not spare the time to make the inquiry; and, indeed, I was not fully aware, at that time, of the very intimate mixture of the supply of the two Water Companies, and the consequently important nature of the desired inquiry (pp. 75—76)

Carrying out the idea required putting together two kinds of data: cholera cases and water supply. The first was easier to come by than the second.

When the cholera returned to London in July of the present year, however, I resolved to spare no exertion which might be necessary to ascertain the exact effect of the water supply on the progress of the epidemic, in the places where all the circumstances were so happily adapted for the inquiry. I was desirous of making the investigation myself, in order that I might have the most satisfactory proof of the truth or fallacy of the doctrine which I had been advocating for 5 years. I had no reason to doubt the correctness of the conclusions I had drawn from the great number of facts already in my possession, but I felt that the circumstances of the cholera-poison passing down the sewers into a great river, and being distributed though miles of pipes, and of so vast importance to the community, that it could not be too rigidly examined, or established on too firm a basis (p. 76).

Snow began to gather data on cholera deaths in the district. The very first results were supportive of his conjecture: of 44 deaths in the district in question, 38 occurred in houses supplied by the Southwark and Vauxhall Company.

As soon as I had ascertained these particulars I communicated them to Dr. Farr, who was much struck with the result, and at his suggestion the Registrars of all the south districts of London were requested to make a return of the water supply of the house in which the attack took place, in all cases of death from cholera. This order was to take place after the 26th August, and I resolved to carry my inquiry down to that date, so that the facts might be ascertained for the whole course of the epidemic...(p. 77).

Determining which water company supplied a given house was not always straightforward. Fortunately, Snow found a chemical test based on the fact that when a solution of silver nitrate is added to water containing chlorides a white cloud of insoluble silver chloride is formed. He found that the water from the two companies differed markedly in chloride content and thus could be easily distinguished.

The inquiry was necessarily attended with a good deal of trouble. There were very few instances in which I could at once get the information I required. Even when the water-rates are paid by the residents, they can seldom remember the name of the Water Company till they have looked for the receipt. In the case of working people who pay weekly rents, the rates are invariably paid by the landlord or his agent, who often lives at a distance, and the residents know nothing about the matter. It would, indeed, have been almost impossible for me to complete the inquiry, if I had not found that I could distinguish the water of the two companies with perfect certainty by a chemical test. The test I employed was founded on the great difference in the chloride of sodium contained in the two kinds of water, at the time I made the inquiry...(pp.77—78).

Table 2
DEATH RATES PER 10,000 HOUSES

	No. of houses	Deaths from cholera	Deaths in each 10,000 houses
Southwark & Vauxhall Co.	40,046	1263	315
Lambeth Company	26,107	98	37
Rest of London	256,423	1422	59

Note: The mortality in the houses supplied by the Southwark & Vauxhall Company was therefore between 8 and 9 times as great as in the houses supplied by the Lambeth Company...(p.86).

The difference in appearance on adding nitrate of silver to the two kinds of water was so great, that they could be at once distinguished without further trouble. Therefore when the resident could not give clear and conclusive evidence about the water Company, I obtained some of the water in a small phial, and wrote the address on the cover, when I could examine it after coming home. The mere appearance of the water generally afforded a very good indication of its source, especially if it was observed as it came in, before it had entered the water-butt or cistern; and the time of its coming in also afforded some evidence of the kind of water, after I had ascertained the hours when the turncocks of both Companies visited any street. These points were, however, not relied on, except as corroborating more decisive proof, such as the chemical test, or the Company's receipt for the rates...(p.78)

It is worth noting how careful Snow was to be sure of the facts here — although he could guess the source of the water from its "mere appearance" he relied on more objective proof of its origin.

Snow now expresses the results of his study in quantitative terms. He notes that the Southwark and Vauxhall Company supplied about 40,000 houses in London during 1853 and the Lambeth Company (drawing its water upstream) about 26,000. In the rest of London, where there were over 250,000 houses, there were more deaths than in the houses supplied by Southwark and Vauxhall — 1422 compared with 1263 — but there were six times as many houses, also. What matters here is not the total number of deaths, but the rate of deaths per house or per family. Put another way, if you live in a house supplied by Southwark and Vauxhall, what are your chances of dying, compared with your chances if you live in a house supplied by another company? Snow expressed the rate in death per 10,000 houses, according to the following formula: rate = (death/number of houses) × 10,000. Table 2 gives the results.

Note in Table 2 that Snow did not apply a statistical test to his data to rule out the possibility that the effect was due to chance. The science of statistics had not yet been developed to the stage where such a test could be performed, and the necessity of such a test was not yet realized. If we wanted to perform such a test on Snow's data we could if we make the reasonable assumption that the houses supplied by the two water companies had, on the average, the same numbers of inhabitants. Note that this assumption would not have been reasonable if the two water companies supplied different districts of London. If the Southwark and Vauxhall houses were crowded dwellings of the poor, the population at risk could have been greater than among the Lambeth houses, and more cholera deaths would have been expected even if the water supply had nothing to do with the disease.

Without performing a statistical test though we can safely conclude the result would have been highly significant just by a glance at the table.

VI. WHAT TO DO? — MEASURES TO PREVENT THE SPREAD OF CHOLERA

The last part of Snow's monograph gives his list of recommended measures for preventing the spread of cholera. His ideas did not win immediate acceptance from his medical contemporaries, who felt that he had made a good case for some influence of polluted water in cholera but continued to believe in effluvia theories as an alternative or contributing cause for a while. In any event, his recommendations on the water supply were adopted, and London was spared any further cholera epidemics.

The measures which are required for the prevention of cholera, and all diseases which are communicated in the same way as cholera, are of a very simple kind. They may be divided into those which may be carried out in the presence of an epidemic, and those which, as they require time, should be taken beforehand.

The measures which should be adopted during the presence of cholera may be enumerated as follows:

1st. The strictest cleanliness should be observed by those about the sick. There should be a hand-basin, water, and towel, in every room where there is a cholera patient, and care should be taken that they are frequently used by the nurse and other attendants, more particularly before touching any food.

2nd. The soiled bed linen and body linen of the patient should be immersed in water as soon as they are removed, until such time as they can be washed, lest the evacuations should become dry, and be wafted about as a fine dust. Articles of bedding and clothing which cannot be washed, should be exposed for some time to a temperature of 212 degrees or upwards.

3rd. Care should be taken that the water employed for drinking and preparing food (whether it come from a pump-well, or be conveyed in pipes) is not contaminated with the contents of cesspools, house-drains, or sewers; or, in the event that water free from suspicion cannot be obtained, it should be well boiled, and if possible, also filtered...

4th. When cholera prevails very much in the neighborhood, all the provisions which are brought into the house should be well washed with clean water and exposed to a temperature of 212 degrees Farenheit; or at least they should undergo one of these processes, and be purified either by water or by fire. By being careful to wash the hands, and taking due precautions with regard to food, I consider that a person may spend his time amongst cholera patients without exposing himself to any danger.

5th. When a case of cholera or other communicable disease appears among living persons in a crowded room, the healthy should be removed to another apartment, where it is practicable, leaving only those who are useful to wait on the sick.

6th. As it would be impossible to clean out coal-pits, and establish privies and lavoratories in them, or even to provide the means of eating a meal with anything like common decency, the time of working should be divided into periods of four hours instead of eight, so that the pitmen might go home to their meals, and be prevented from taking food into the mines.

7th. The communicability of cholera ought not to be disguised from the people, under the idea that the knowledge of it would cause a panic, or occasion the sick to be deserted.

The measures which can be taken beforehand to provide against cholera and other epidemic diseases, which are communicated in a similar manner, are:

8th. To effect good and perfect drainage.

9th. To provide an ample supply of water free from contamination with the contents of sewers, cesspools, and house-drains, or the refuse of people who navigate the rivers.

10th. To provide model lodging-houses for the vagrant class, and sufficient house room for the poor generally...

11th. To inculcate habits of personal and domestic cleanliness among the people everywhere.

12th. Some attention should be undoubtedly directed to persons, and especially ships, arriving from infected places, in order to segregate the sick from the healthy. In the instance of cholera, the supervision would generally not require to be of long duration...

I feel confident, however, that by attending to the above-mentioned precautions, which I consider to be based on a correct knowledge of the cause of cholera, this disease may be rendered extremely rare, if indeed it may not be altogether banished from civilized countries. And the diminution of mortality ought not to stop with cholera...(p. 133)

VII. EXERCISE FOR STUDENTS

1.   Using only information available to Snow (as given in this chapter) design an appropriate case-control study to identify for the causal factors in cholera. How would you test the significance of your results?
2.   Describe the factors observed by Snow in terms of accepted criteria for causality in epidemiology as currently practiced.

REFERENCES

1. **Rosen, G.**, *A History of Public Health*, M.D. Publ., New York, 1958.
2. **Rosenberg, C. E.**, *The Cholera Years*, University of Chicago Press, Chicago, Ill., 1968.
3. **Chadwick, E.**, *The Sanitary Condition of the Labouring Population of Great Britain*, reprinted by Edinburgh University Press, Edinburgh, 1965.
4. **de Morgan, A.**, *Budget of Paradoxes*, Vol. 1, Open Court Publ., La Salle, Ill., reprinted 1915.
5. **Snow, J.**, On the mode of communication of cholera, republished in *Snow on Cholera*, The Commonwealth Fund, New York, 1936.

Chapter 3*

CHOLERA OUTBREAK IN JERUSALEM 1970, REVISITED: THE EVIDENCE FOR TRANSMISSION BY WASTEWATER IRRIGATED VEGETABLES

Badri Fattal, Perez Yekutiel, and Hillel I. Shuval

TABLE OF CONTENTS

*   This study was carried out as part of a project commissioned by the World Bank as executing agency for the UNDP Integrated Resource Recovery Project (GLO/80/004). The views presented are the authors' and should not be attributed to the World Bank, to its affiliated organizations, or to the United Nations Development Program.

I. INTRODUCTION

Cholera attracted worldwide attention when the seventh pandemic began in 1961 as the causative agent, the El Tor biotype, spread north and east from its endemic focus in Indonesia to the Indian subcontinent, reaching as far as the Middle East, Eastern Europe, and Western Africa by 1970.[1] As described by Cohen, et al.,[2] unsubstantiated reports reached the Israeli public health authorities of cholera cases in Middle Eastern countries from which numerous visitors had entered Israel in the summer of 1970.

The first cases of cholera were detected in Jerusalem on August 20, 1970. As the Jerusalem outbreak spread, several features became clear: the sporadic, seemingly unconnected appearance of cases in widely separated areas and among various ethnic groups with little likelihood of personal contact; the fact that few co-primary or secondary cases were observed among close contacts, such as family or co-workers; and the lack of spread outside of the Jerusalem area, although it was open to traffic from all areas of Israel throughout the outbreak.

All these facts indicated that there was a common source of infection such as food, water, or wastewater. Mechanical spread via flies was unlikely, since the flies had no access to feces. Open latrines were not in use and, as can be seen from the map, the Jerusalem sewage system is a closed network, therefore, cholera transmission by flies can be discounted.[3] The Jerusalem water supply is drawn from deep and chlorinated wells, and routine bacteriological tests for coliform bacteria indicated that the water was of high bacteriological quality, thus, the water was ruled out. All milk and dairy products were pasteurized and under strict microbial quality control and were thus, also ruled out as a possible common source.

Cohen et al.[2] stated that "...it seems likely that after the Jerusalem sewage had become contaminated with vibrios from cases, mild cases, and carriers, vegetables irrigated with that sewage in surrounding villages constituted a major secondary vehicle for the spread of the infection". Feachem[1] in his review of Cohen et al.[2] wrote "No solid evidence is presented to support this hypothesis [i.e., vegetables irrigated with wastewater] but it is plausible. When scattered cases of cholera occur widely in a community using fully treated water and not consuming raw seafood, the number of possible vehicles of transmission is very small. It is unfortunate that studies were not conducted during the outbreak to prove or disprove the vegetable hypothesis. Subsequent laboratory studies showed that *Vibrio cholerae* (El Tor) could survive sufficiently long in sewage and soil, and on vegetables to make vegetable-borne cholera in Jerusalem theoretically possible".

As the subject of transmission of infectious diseases by crops fertilized by night soil or irrigated with wastewater is of special interest to developing countries, and in light of planned wastewater reuse programs for irrigation, it is worthwhile to re-examine the 1970 outbreak to verify, if possible, the mode of transmission, despite the fact that the Jerusalem outbreak occurred more than a decade ago. Although that outbreak has been reported in numerous publications, most of them were in local Israeli literature not fully available throughout the world.

The authors, who had participated actively in investigating the environmental aspects of this cholera outbreak, decided that it would be useful to reanalyze the published data and present additional previously unpublished information to further evaluate the hypothesis of transmission via vegetables irrigated with wastewater, which was suggested by Cohen et al.[2]

II. DESCRIPTION OF THE JERUSALEM CHOLERA OUTBREAK

The first three cases of cholera in Jerusalem appeared simultaneously in three different locations and in widely disparate social groups. Within 12 hr of admission on August 20th, 1970, of a 30-year-old woman living in a Jerusalem suburb to hospital, confirmation was

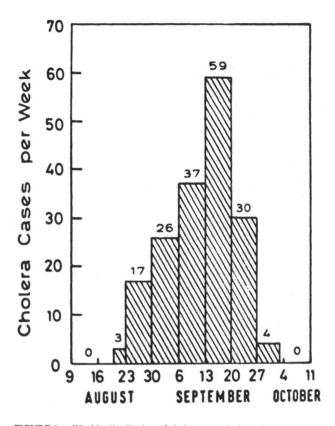

FIGURE 1. Weekly distribution of cholera cases in Jerusalem, August—
October 1970 (T = 176).

obtained of the first case of cholera.[2,4] As demonstrated by serological testing, the outbreak
consisted of two serotypes of *V. cholerae:* El Tor, Inaba and Ogawa. Within several weeks,
the cholera had spread rapidly, and eventually, 258 acute cases (176 in Jerusalem and 82
cases in nearby villages) were confirmed. Figure 1 presents the weekly numbers of the 176
Jerusalem cases, from Sunday to Saturday. It can be seen that cholera started with three
cases in the 1st week, increased to 17 in the 2nd week, then 26, followed by 37, and reached
a peak in the 5th week of 59 cases. It then dropped to 30 cases and the 7th and the final
week only had 4 cases.

This epidemic was investigated from several angles in order to clarify the mode of
transmission. During the Jerusalem outbreak 4,560 specimens of rectal swabs from asymptomatic contacts were examined. Fewer than 3% were positive for *V. cholerae.* There were
1073 stool specimens from the population in the affected area, and 0.6% were found to be
positive. Of the 1359 stools examined among food employees in Jerusalem only 0.3% were
positive.[5] These low rates indicate that the spread of cholera in this outbreak was not generally
via direct contact with patients or food handlers.

In a substantial number of cases, intensive individual epidemiological case investigations
were carried out. These investigations did not reveal alternative sources of infection (such
as contact with other cases or their immediate contacts, with the exception of very few co-
primary and secondary close contact cases, specified food stores or eating places, or other
specified food articles), but in a number of instances definite information was obtained on
purchase of raw-consumed fresh vegetables and herbs (e.g., parsley) from village peddlers.
In some instances the family member who had fallen ill with cholera was the only one who
had consumed these products.

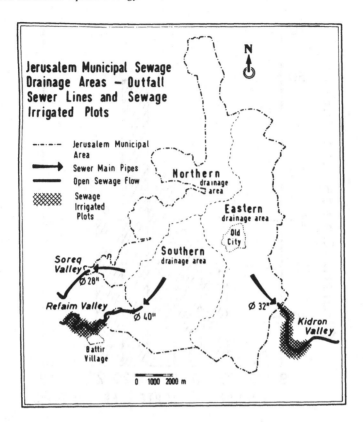

FIGURE 2.   Jerusalem sewage system.

III. COMMUNITY BACKGROUND AND JERUSALEM SEWAGE SYSTEM

A short digression is in order, to describe the community background of the outbreak. As used in this report, the "Jerusalem area" corresponds to the city of Jerusalem and the surrounding rural villages. The city itself, in 1970, had a population of about 260,000, ranging widely in ethnic origin, socio-economic level, and hygienic habits. Except in several small villages on the city outskirts, sanitary facilities included a sanitary piped chlorinated groundwater supply and water closets in essentially all homes, connected to a central sewage system draining into a number of valleys outside of the city limits, as illustrated in Figure 2 (map). There were three major routes of untreated sewage flow after leaving the city in closed sewers, leading to open canals. The eastern route ran through the Kidron Valley to the Dead Sea, and the northern route was through the Soreq Valley; the southern path ran through the Refaim Valley. The latter two joined to flow towards the Mediterranean Sea. The Refaim Valley route passed the village of Battir, among others. In the villages near the Kidron and Refaim Valleys, vegetable crops were irrigated by flood irrigation methods with untreated wastewater drawn from the open canals, as shown in Figure 2. As stated in the Israel Ministry of Health regulations regarding "Conditions for Granting Permits for Wastewater Utilization in Agricultural Irrigation",[6] written consent of the Ministry of Health is clearly required by law and irrigation of vegetable crops eaten raw is forbidden. It must be noted here that the villages in the Kidron Valley and Battir were in the area administered by the Hashemite Kingdom of Jordan prior to the Six Day War of 1967. From a legal point of view Jordanian law still applied in those Israel-administered areas, thus, the strict Israel Ministry of Health regulations as to the use of wastewater for irrigation of crops were not legally in force in those villages. Since 1967, when the City of Jerusalem was united, a minor proportion of the vegetables supplied to Jerusalem were provided by these villages, usually by means of door-to-door peddlers who circulated in the city, although a portion of the vegetables were also marketed through the central municipal vegetable market and other vegetables probably reached restaurants. Much of this produce was consumed on the same day, and if not immediately consumed was usually refrigerated, which tended to preserve the bacteria.

Various studies have been carried out on the viability of cholera pathogens in seawater, feces and night-soil, sewage sludge, wastewater, sweat on surfaces, in soil, and on food or crops.[7-10] Gerichter et al.[7] studied the viability of cholera vibrios (El Tor) strains isolated in the Jerusalem outbreak on a variety of vegetables and survival times ranged up to 24 hr on parsley, 30 hr on tomatoes and carrots, nearly 48 hr on cucumbers, peppers and okra, and 2 to 3 days on lettuce at room temperature during the summer. The produce was inoculated with vibrios and cholera phages isolated from stools of patients who were ill with cholera during the summer 1970 outbreak in Jerusalem. Although isolatable from experimentally contaminated vegetables for up to $2^1/_2$ days, the vibrios die rapidly on dry vegetables or soil or in sunlight, underscoring several important points: firstly, flooding of wastewater irrigated fields aids in maintaining viable vibrios (up to 10 days in moist soil[7]); secondly, as sunlight, lack of refrigeration and desiccation decrease viability of the cholera pathogen, it is important that during epidemiological study of cholera, the vegetables examined as possible vectors of transmission be transferred under optimal conditions as fresh samples, with moisture preserved, in order to enhance reliability of the bacteriological tests. During the very first days of the outbreak in August to October 1970, vegetables from suspect sources were not cultured for cholera vibrios or other indicators of infections such as cholera phages.[11-14] The first tests of wastewater irrigated soil and vegetables were initiated by the Environmental Health Laboratory a few weeks after the start of the epidemic. Our positive findings of cholerae vibrio in the wastewater and wastewater irrigated soil, later confirmed in extensive tests by the Ministry of Health, provided more than circumstantial evidence pointing strongly to the hypothesis that vegetables irrigated with wastewater played a central role in the transmission of cholera.

The proposed sequence during the Jerusalem cholera outbreak in summer 1970 was as follows:[1,2] cholera was introduced into villages outside of Jerusalem or to the city itself by travelers from other Middle Eastern countries in which cholera was endemic at the time. During the summer of 1970 some 100,000 persons visited the area from Middle Eastern countries. Many persons from these villages come to work daily in all areas of Jerusalem and a few subclinical or clinical cases were sufficient to contaminate the city sewage with cholera vibrios from the feces of infected individuals. This wastewater was used, as mentioned above, to irrigate vegetable fields, thus contaminating vegetables and establishing a cycle of infection and widespread transmission as illustrated in Figure 3. We have established that one clinical case of cholera can excrete 10^{10} to 10^{12} cholerae vibrio per day to the sewage system, which is sufficient to cause contamination at the infectious dose level in vegetables irrigated with fresh raw sewage from that neighborhood.

IV. TEST FINDINGS FOR CHOLERA INFECTION

A. Bacteriological Findings

During the 6-week outbreak, about 10,000 stool samples or swabs were examined for *Vibrio cholerae*[4] (2762 suspected cases and 7010 controls and survey subjects). Bacteriological confirmation was obtained in a total of 258 cases, as well as for 150 asymptomatic contacts. The outbreak did increase consultations with primary-care physicians for all types of gastrointestinal problems, leading to higher morbidity figures for *Salmonella* and *Shigella* due to the increased number of stool samples processed in bacteriological laboratories,[15] an

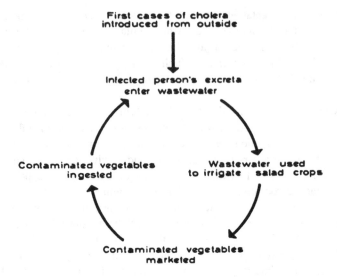

FIGURE 3.   The cycle of transmission of *V. cholerae* from one infected person to another via vegetables irrigated with wastewater.

artifact of the outbreak which skewed the morbidity figures for all gastrointestinal disease that year. Laboratory methods used to diagnose cholera included characterization of *V. cholerae* based on serological and biochemical examinations, as well as phage typing, as described by Gerichter et al.[4]

B. Serological Findings

A serological survey was conducted on sera collected from bacteriologically confirmed cholera cases (2 to 3 months after the 1970 endemic), as well as from pre-epidemic controls and from non-affected controls. The sera were examined using the vibriocidal antibody microtest as well as a parallel microagglutination test. The rates of inapparent infection, as reflected by these serological findings, were about 8% in the Jerusalem population and about 57% in the village of Battir, where the fields were irrigated with wastewater from the city. This village also had the highest incidence of cholera cases. Cholera vibrios and phages were successfully isolated from vegetables grown in its fields. The inapparent rate of infection in Battir was the same as in a number of other villages and about sevenfold the rate in Jerusalem. It may be concluded that in Battir, over half the population was contaminated with *V. cholera* but only about 1 in 110 persons infected actually developed clinical symptoms.

C. Isolation of *V. cholerae* and Cholera Phage from Wastewater

During the outbreak the Environmental Health Laboratory initiated a program of sampling wastewater and wastewater irrigated soil.[5] Our detection of vibrios in the wastewater from all areas of the city and from soil irrigated with wastewater in the village of Battir provided the first firm evidence of the vegetable route of infection. Later the sampling program was continued and expanded by the Ministry of Health. A total of 143 wastewater samples were taken by the gauze pad method and by the grab sample method for cholera isolation. Of these, 72 were taken during the outbreak (September 6th to October 3rd, 1971), and 71 were taken during the following months (January 1st, 1971 to June 1st, 1971). For cholera phage isolation, a total of 168 samples were examined and of these, 29 were taken during the outbreak, 57 were 2 months afterwards, i.e., at the end of 1970, and another 82 were taken over the next 6 months. Table 1 summarizes the results of *V. cholerae* and cholera phages isolated from wastewater effluent based on the work of Gerichter et al.[5]

From this table it can be seen that during the cholera outbreak, wastewater samples were

Table 1
SUMMARY OF MICROBIOLOGICAL TESTING DURING THE CHOLERA OUTBREAK[a]

Test	Period of testing	Days testing began after outbreak	Results
V. cholerae in Jerusalem wastewater	During outbreak Sept. 6 to Oct. 3, 1970	17	72 samples 13 positive (18%)
	After outbreak Jan. to June 1971		71 samples 0 positive
Cholera phages in Jerusalem wastewater	During outbreak Sept. 9 to Oct. 3, 1970	20	29 samples 19 positive (66%)
	After outbreak Oct. 4 to Nov. 11, 1970		57 samples 35 positive (61%)
	Nov. 12, 1970 to June 1971		82 samples 0 positive (0%)
V. cholerae on vegetables irrigated with Jerusalem wastewater	During outbreak Sept. 8 to Dec. 31, 1970	19	194 samples 1 positive (0.5%) (on parsley from Battir)
	After outbreak Jan. to June 1971		150 samples 0 positive (0%)
Cholera phages on vegetables irrigated with Jerusalem wastewater	During outbreak Sept. 8 to Dec. 31, 1970	19	194 samples 6 positive (3.1%) (4 from Battir)
	After outbreak Jan. to June 1971		150 samples 0 positive (0%)
V. cholerae and phage in food products	Not mentioned	Not mentioned	48 milk and milk products, all negative (0%)

[a] Based on Gerichter et al.[5]

positive for *V. cholerae* (18%) and no positive samples were detected when taken after the outbreak, i.e., between January 1971 and June 1971. On the other hand, cholera phages were frequently found during the outbreak (66%) and 4 weeks after the end of the outbreak (61%), and none were found afterwards.

D. Isolation of *V. cholerae* and Cholera Phage from Vegetables

The isolation process of *V. cholerae* and cholera phage from vegetables was initiated on September 8th, that is 19 days after the start of the outbreak.[5] During the outbreak and for 8 months thereafter, 344 samples were tested of the following vegetables: parsley, cucumber, zucchini, radish, eggplant, tomato, beans, beet leaves, green pepper, onion, carrots, mint, pumpkin, and *Hibiscus esculentus*.

The results of *V. cholerae* and cholera phages tests of vegetables are found in Table 1. From this table it can be seen that cholera vibrio was isolated only once from vegetables vs. six isolations of cholera phages.

The positive cholera phages were isolated on September 24th from a specimen of parsley in the Jerusalem market and the second on September 25th from a tomato stored in the refrigerator of one of the confirmed acute cases. The other four specimens for cholera phages which were isolated on October 18th were collected at Battir, where some wastewater irrigation was apparently still going on. The positive sample of *V. cholerae* El Tor, Inaba, phage type 6, isolated on October 18, 1970 was from a specimen of parsley grown in the village of Battir.

No positive results were found in milk products and the water supply was well-chlorinated with no signs of pollution.[5]

V. CONTROL MEASURES

Based on our detection of vibrio in the wastewater streams being used for vegetable irrigation and in wastewater irrigated soil at Battir and prior to the first confirmed isolation of vibrios or phages on the vegetables themselves, the Ministry of Health authorities decided to impound and destroy all vegetables irrigated with Jerusalem wastewater as well as to forbid marketing them. This was carried out on or about the 15 to 20th of September, 1970 and completed within a few days, although a few farmers obviously renewed the practice some time later. As illustrated in Figure 1, there was a drastic and significant drop in the number of cholera cases already in the first week after these measures were taken, with total disappearance 2 weeks later. No cholera vibrios were detected in wastewater after the last clinical case.

VI. EFFECT OF OUTBREAK ON ENVIRONMENTAL HEALTH REGULATIONS

Although the reports published in the scientific literature at the time of the outbreak did not emphasize the importance of the wastewater irrigated vegetables as the major and massive route of secondary transmission, the Israel public health community in general and the officials of the Ministry of Health, in particular, fully appreciated the serious environmental implications. Israel, as a country facing severe water shortages, was at that time already promoting wastewater reuse for agricultural irrigation. Some 200 wastewater irrigation projects were already operating. As a result of the Jerusalem cholera outbreak, the Ministry of Health appointed a committee which recommended several modifications in the already existing wastewater irrigation regulations making them even stricter than before. While wastewater irrigations of vegetables eaten raw had been forbidden by the previous regulations, the new ones forbad irrigation of crops for direct human consumption even when generally eaten cooked. This was done to avoid the risk of contaminated vegetable entering kitchens and contaminating surfaces, utensils, and hands prior to the cooking of the vegetables.

A second major impact of the epidemic was a government decision to make a major investment in improving municipal wastewater disposal and treatment works throughout Israel. A national sewage plan was developed, partially funded by the World Bank, which has resulted in major environmental improvement in wastewater treatment and disposal. In Jerusalem a wastewater treatment plant has been built for a portion of the city and the open flow of sewage in the dry river beds has been partially replaced by closed sanitary sewers.

VII. CHOLERA OUTBREAK IN GAZA

In this connection it is worth mentioning the outbreak of cholera which occurred in Gaza during the period of November 16 to December 12, 1970. This outbreak was also in a limited area and only 260 acute cases were reported for the period of 5 weeks. *V. cholerae* and cholera phages were found in wastewater and vegetables used by affected population groups. Gerichter et al.[5] stated that the experience gained in finding *V. cholerae* in vegetables in Jerusalem, and the fact that at the beginning of the Gaza outbreak, the search for *V. cholera* in vegetables started immediately, may have played a significant role in the early positive results obtained in Gaza. It should be pointed out that the drinking water in Gaza, compared to Jerusalem, was found to be contaminated with coliform as well as with cholera phage. This indicates the possibility that the outbreak in Gaza was also spread by the drinking water route, thus it is difficult to determine the relative role of transmission of *V. cholera* via vegetables irrigated with wastewater. Therefore, this outbreak is less relevant to the subject addressed in this article.

VIII. CONCLUSIONS

Our traditional conception of cholera epidemiology has undergone considerable modification since John Snow described the main features of cholera transmission and control, as related in Feachem's historical review[16] and in Goldstein and Goldstein.[17] Although the relationship between cholera transmission and water is central to understanding the spread of the disease, the assumption that cholera is a mainly waterborne infection was erroneously derived from Snow, who had also discussed the fecal-oral route and the risk of foodborne transmission.

The findings and conclusions of this study can be summarized as follows:

1. The rapid simultaneous appearance of numerous cases of cholera in widely spread areas of Jerusalem which on investigation showed little or no evidence of contact between the cases of cholera indicated that a common source epidemic was occurring.
2. The fact that the city was supplied with a chlorinated water supply from deep protected sanitary wells that met rigid bacterial standards by frequent monitoring for coliform bacteria ruled out water as a possible common source.
3. Milk and dairy products supplied to the city were also ruled out since they were all pasteurized and produced in plants of generally high sanitary standards and underwent routine testing.
4. No other food product was considered at first as a possible common source. Epidemiological investigations did not reveal alternative food-related sources of infection (e.g., eating places or other specified food articles); in cases of cholera in west Jerusalem, a history of purchase of fresh vegetables from peddlers was elicited from family members.
5. Mechanism of spread by flies was ruled out as there are no open latrines within the city.
6. The detection of cholera vibrios and cholera phages in the main outfall sewers discharging wastewater to a number of dry river beds adjacent to the city where unauthorized irrigation of salad crops was being carried out led to the hypothesis that these vegetables, widely marketed by peddlers in all neighborhoods within the city, were the main secondary mode of spread of the disease after the first initial cases were introduced to the city by visitors from other Middle Eastern countries where cholera epidemics were known to exist.
7. The hypothesis of transmission by wastewater irrigated vegetables was further strengthened by the isolations of vibrios and phages in wastewater irrigated soil and vegetables in the field as well as vegetables purchased in the market and refrigerated in the home of a clinical case.
8. The fact that the epidemic came to a rapid and dramatic end within a short period after the wastewater irrigated vegetables were confiscated and the crops still growing in the field were destroyed provided strong circumstantial evidence that the wastewater irrigated vegetables were indeed the main, if not sole, secondary mode of transmission of cholera in widely separated areas of the city after introduction of the first cases from neighboring countries.

The main conclusion of this re-evaluation of the 1970 cholera epidemic in Jerusalem is that cholera vibrios can be present in municipal wastewater in high concentrations and can survive in the wastewater and in wastewater irrigated soil and vegetables for sufficiently long periods and in high enough concentrations to infect the vegetables with minimal infectious doses capable of causing clinical cases of cholera in persons consuming such vegetables. We have estimated that a single clinical case of cholera can excrete 10^{12} vibrios per day and that this one case can lead to a vibrio concentration of some $10^6/100$ mℓ in the wastewater of a community having a population of 10,000 persons (assuming 100 ℓ per cap per day of sewage flow). Irrigation of vegetables eaten raw, particularly salad crops, with such wastewater can thus lead to the transmission of cholera to a portion of the population consuming such vegetables. Therefore, the use of raw sewage, or even partially treated wastewater, for the irrigation of vegetables eaten raw can be one of the modes of transmission of cholera in endemic areas or during epidemic periods. This is particularly important in areas where other classical routes of transmission of the disease are closed.

This analysis is in no way intended to imply that wastewater irrigated vegetables are the main mode of cholera transmission since in most areas with insufficient and/or contaminated water supply, inadequate disposal of human excreta and low levels of personal hygiene, all of the other well-known routes of cholera transmission, particularly personal contact, are dominant.

ACKNOWLEDGMENTS

This chapter is dedicated to the memory of Dr. J. Cohen, who guided the research through all its stages. We are also grateful for the assistance of Drs. Schwartz, Gerichter, and Sechter who have continued in their efforts regarding the environmental aspects, and to O. Grafstein and S. Stambler for editorial assistance.

REFERENCES

1. **Feachem, R. G.,** Environmental aspects of cholera epidemiology. I. A review of selected reports of endemic and epidemic situations during 1961—1980, *Trop. Dis. Bull.,* 78(8), 675, 1981.
2. **Cohen, J., Schwartz, T., Klasmer, R., Pridan, D., Ghalayini, H., and Davies, A. M.,** Epidemiological aspects of cholera El Tor outbreak in a non-endemic area, *Lancet,* July 10, 86, 1971.
3. **Dishon, T.,** Factors Affecting the Transmission of Pathogenic Bacteria by Insects *C. vibrio comma* and *Salmonella paratyphii* B in the House Fly *Musca vicina,* Ph.D. thesis, Hebrew University, Jerusalem, 1955 (in Hebrew; abstract in English).
4. **Gerichter, C. B., Sechter, I., and Cahan, D.,** Laboratory diagnosis of cholera during the Jerusalem outbreak August-September 1970, *Isr. J. Med. Sci.,* 8(4), 531, 1972.
5. **Gerichter, C. B., Sechter, I., Cahan, D., and Gavish, A.,** Laboratory investigations during the cholera outbreak in Jerusalem and Gaza 1970, *Bri'ut Hatzibur (Public Health),* Hebrew with English abstract, 3, 26, 1971.
6. Ministry of Health, Conditions for Granting Permits for Wastewater Utilization in Agricultural Irrigation, (revised) July 1979 (was valid also in 1970).
7. **Gerichter, C. B., Sechter, I., Gavish, A., and Cahan, D.,** Viability of *Vibrio cholerae* biotype El Tor *and of cholera phage on vegetables, Isr. J. Med. Sci.,* 11(9), 889, 1975.
8. **Feachem, R., Miller, C., and Drasar, B.,** Environmental aspects of cholera epidemiology. II. Occurrence and survival of *Vibrio cholerae* in the environment, *Trop. Dis. Bull.,* 78(10), 855, 1981.
9. **Hood, M. A. and Ness, G. E.,** Survival of *Vibrio cholerae* and *Escherichia coli* in estuarine waters and sediments, *Appl. Environ. Micriobiol.,* 43(3), 578, 1982.
10. **Sechter, I., Gerichter, C. B., and Cahan, D.,** Method for detecting small numbers of *Vibrio cholerae* in very polluted substrates, *Appl. Microbiol.,* 29(6), 814, 1975.
11. **Sechter, I., Gerichter, C. B., and Cahan, D.,** An Indirect Method for Detecting the Presence of *Vibrio cholerae,* presented at the Int. Cong. Bacteriol., Jerusalem, September 1973.

12. **Sechter, I., Cahan, D., Rogol, M. and Gerichter, C. B.,** Rapid confirmation of cholera with the aid of specific bacteriophages, *Isr. J. Med. Sci.,* 14(7), 816, 1978.

13. **Rogol, M., Sechter, I., and Gerichter, C. B.,** Technique for detecting scarce *Vibrio cholerae* colonies on a solid medium, *Isr. J. Med. Sci.,* 16(1), 76, 1980.

14. **Rogol, M., Sechter, I., and Gerichter, C. B.,** Identification of some auxotrophic strains of *Vibrio cholerae, Isr. J. Med. Sci.,* 16(1), 75, 1980.

15. **Cohen, J., Ever Hadani, S., Shallom, Y., and Eden, T.,** The cholera outbreak in Jerusalem 1970 as part of the intestinal infections problem in Israel, *Bri'ut Hatzibur (Public Health),* Hebrew with English Abstract, 3, 3, 1971.

16. **Feachem, R. G.,** Environmental aspects of cholera epidemiology. III. Transmission and control, *Trop. Dis. Bull.,* 79(1), 1, 1982.

17. **Goldstein, M. and Goldstein, I. F.,** Snow on cholera, in *How We Know,* Plenum Press, New York, 1980, chap. 3.

18. **Gerichter, C. B., Sechter, I., Cohen, J., and Davies, A. M.,** A serological survey for cholera antibodies in the population of Jerusalem and surroundings, *Isr. J. Med. Sci.,* 9(8), 980, 1973.

Chapter 4

CHOLERA TODAY

John R. Goldsmith

Four historic events have transformed cholera from the devastating and dreaded disease of the mid-19th Century to a treatable and preventable condition. Cholera seems to have been tamed by man.

Two of the events are recounted in the preceding chapters. The recognition of the transmissibility of cholera in drinking water by Snow initiated the modern era of supply of safe drinking water to modern cities, a supply which makes possible much of the urbanization of the 20th Century; for what was true for cholera was also true for typhoid and other diarrheal diseases; they too can be spread by fecal contamination of drinking water, and modern drinking water supplies protect against their spread, not only by providing a safe supply, but also by providing a readily available source of water for handwashing and sanitation.

Prior to the second of the historic changes, presented in Chapter 3, Section II., we had and used two relatively effective tools for management of cholera outbreaks, immunization, and antibiotic therapy. The difficulties with these tools were that immunization was not 100% effective and once a population was immunized, either naturally or by shots, the immunity did not, for some reason, last very long. So before 1970, travelers to areas where cholera was prevalent had to have a new series of shots for each trip, and when a cholera outbreak occurred, huge amounts of cholera vaccine were rushed to the site. Even so, fear and ignorance led to the occurrence of cholera being the excuse for cutting off mail and even telegraphic communication to the affected area! In areas where cholera continued to occur, effective treatment with antibiotics was not able to stop the epidemic, nor quell the panic. The management of the 1970 Jerusalem epidemic by tracing it to its source and breaking off the cycle of fecal-oral transmission was the first to control a cholera epidemic without use of vaccine (immunization). This application of epidemiology was the forerunner of the altered epidemiological strategy for the control and apparent eradication of smallpox, one of the most astonishing accomplishments of modern epidemiology.

The third historic event was a natural change in the strain of the cholera vibrio which causes the disease. In 1961, the El Tor strain of cholera emerged in what Fattal et al. correctly called the seventh pandemic — the seventh worldwide epidemic of cholera. This strain is relatively mild compared to those in previous pandemics, and this attribute has given us the opportunity to develop and to apply the fourth historic change in our relationships with cholera. We cannot know whether the next, the eighth pandemic, may not be much more virulent, and the ecological balance may once again swing back to cholera, and against man.

The fourth historic event is the development of simple household methods for treatment of cholera and other diarrheal disease, known as "Oral Rehydration Therapy", ORT. Cholera threatens life because its toxin affects water retention in the gastrointestinal tract, and the loss of water in the stools leads to dehydration and vascular collapse. Intravenous injection of fluids was the most urgent and common procedure used to treat hospitalized patients. Cholera patients are thirsty and water is easily retained by mouth, but water alone is not sufficient, for salts are lost as well in the watery stools. But a mixture of salt, sugar, and water is easily retained by mouth and it prevents dehydration and collapse until the body's own defenses with or without help from antibiotics can allow recovery from the toxic effects of cholera. As with a good water supply, this strategy is effective for other dysenteric diseases as well. Today in the areas still threatened by cholera, a mass education program has put into the hands of ordinary people the means for their own treatment of this former menace. Doctors, hospitals, and even clinics are less critical which is fortunate as they have been in short supply in such areas. Primary care for cholera and diarrheal diseases is ORT, a modest packet of salt and sugar that is to be mixed with water and given to the sick person with diarrhea, without waiting for the doctor or nurse.

This seventh pandemic of cholera with the El Tor strain has spared the Western Hemisphere during the early stages, at least. Between 1973 and 1981 only 30 cases were identified in the U.S. and all were on the Gulf Coast in Louisiana and Texas. On June 15, 1983 a tourist returning from the Island of Cancun off the coast of the Yucatan peninsula of Mexico was diagnosed as having cholera on the basis of bacteriological examination. This is the first case apparently acquired in a Western Hemisphere country outside of the U.S. since 1900! Indeed only 10 cases of cholera in U.S. travelers have been reported during the first 20 years of this seventh pandemic; this leads the "Morbidity and Mortality Weekly Reporter" the U.S. Center for Disease Control's widely read bulletin to conclude that the risk to tourists going to the Yucatan area should be very slight.

Today, we feel that we know how to prevent and treat cholera, what is the cause and mechanism of transmission, and we have put into the hands of ordinary people the tools for protecting themselves from the most severe impact should cholera strike. The lessons we have had to learn have turned out to be most useful in preventing and treating diarrheal and other diseases. Community epidemiological studies in London and in Jerusalem have been the key to our success.

Part B

Chapter 5*

THE 20-MINUTE DISASTER; HYDROGEN SULFIDE SPILL AT POZA RICA

John R. Goldsmith

TABLE OF CONTENTS

*   Names of characters in this chapter are fictitious, as are some of the personalized comments in Section II.

I. ACCOUNT OF A VICTIM

Our neighborhood is near the plant, which is probably why so many of my neighbors and friends were stricken. We moved there when the plant opened and our house was better built than many of my neighbors, because I was able to get some lumber from the port where my brother-in-law works. Possibly that is the reason why we all survived.

We have a nice climate and for as long as my grandfather could remember we had grown our own food; my father raised bananas, and his neighbors raised pineapples, coconuts and papayas for the market which gave us a chance to have a little money and buy a radio and some furniture. Our region is blessed, according to some people, because oil and gas have been found by the drillers, and work in the oil and gas fields and the plant gives us an even better chance to earn money and save for our daughter's wedding and my father's funeral — may it be many years away! There is a lot of construction work going on in the city and in the plant. Our 16-year-old son is getting to be a good carpenter, and our son-in-law makes good money as a pipefitter. The younger children take care of the chickens, geese, and ducks. All eight of us were at home that awful morning, and because it had been my birthday and we'd had a nice party, we may have slept later than did some of our neighbors. Rosa, my wife was already up and making coffee, when Manuel our daughter's husband woke up and shouted "Gas is leaking" "Cover your face". I thought at first he was having a nightmare, but luckily all of the children ducked under covers, and then I began to hear the screams from all around us. I started to go out, but Manuel, Flora, and Rosa grabbed me and shrieked together "Donna mia, help us, protect us!" Manuel told me later that he recognized the smell from the work he'd been doing at the plant, and they'd been told to hold your breath and let the wind blow it away, but never, never to run or take a big breath, because that would make it harder to save yourself. He'd told all this to Flora our big girl just the night before and Flora and Rosa had discussed it when cleaning up after the party. So despite the shrieks and groans, and two of the children being sick, we covered our faces with towels, and I turned on the radio just in time to get the 5 o'clock news. I remember seeing the superintendent's car race toward the plant, just before but with Manuel's urging I stayed in the house, although I was beginning to get sick to my stomach, and was beginning to think bed would be a better place for me than the plant anyhow. I don't remember how long we stayed there, but Flora kept calling that from the window she could see people lying near the road, while others were heading toward the hospital. Some ambulances were coming back and forth on the road, and then I remember the foreman banging on the door and calling for me and Manuel to help shut the new unit, and he was wearing a gas mask and had one for each of us. I was glad to put it on and only then ventured to take a deep breath and my stomach began to settle down. He warned the rest of the family to stay in the house and keep the sheets and towels wet and over our faces.

It seemed only a few minutes later that we had shut the valves and the sun was out and the fog burned away, but already 13 people were dead and the hospital was filled and the size of the disaster began to be felt. I myself lost a brother and sister-in-law, 2 little nieces, 11 of our 31 chickens, 5 of 8 puppies, 2 geese, 3 ducks, our little canary bird, and later on the pig died, but my dear wife and children were spared.

II. ACCOUNT OF THE SUPERINTENDENT

Superintendent Marino was awakened at his hillside home by a telephone call from the rivergate guard. He called the foreman, picked him up and drove to the southgate safety shack to get gas masks when he saw that the flare was out. The foreman rounded up the emergency crew and together they shut the valves. Two weeks later his final report to headquarters contained the following information:

Our plants together have a capacity of 116 MCFD (million cubic feet of gas per day) and

5000 barrels of hydrocarbon liquids. The gas is "sour", that is it contains a high proportion of hydrogen sulfide, 3.14% to be exact, along with 15% carbon dioxide and 0.1% oxygen. To meet the growing demand for industrial sulfur we had been building a sulfur recovery unit. Girbotol units for stripping the hydrogen sulfide, using monoethanolamine were being completed, and half of them were put into operation on November 21. From the stripping columns the gas was to go to reaction chambers based on the Klaus process where a third of the in-coming hydrogen sulfide gas is burned to sulfur dioxide, which is combined catalytically with the remainder to form elemental sulfur and water. Since the reactors were not then completed, the hydrogen-sulfide rich gas went to the flare, 90 ft up in the air, where it was to be burned. The flare was capable of handling 11 MCFD of gas which averaged 81% carbon dioxide, 16% hydrogen sulfide, and 3% hydrocarbons and water. To assure that the mixture was combustible, auxiliary gas lines delivered sulfur-free natural gas to the combustion chamber at the top of the flare at the rate of 0.288 MCFD. The flare had pilot lights and an automatic electrical ignition system. When the scrubbers were first put into operation, there was some difficulty in establishing a steady flow of gas, and some of the monoethanolamine overflowed into the lines supplying combustible gas to the pilot lights and flare. Therefore the flare was reduced by half its normal height of 30 ft, the absorption units were shut down, the lines drained, and the flare relighted. It appeared to burn satisfactorily thereafter.

For the half of the plant served by the strippers we processed gas at 40 MCFD, which was but 2/3 of the 60 MCFD capacity, until 2 am on November 24, when, as scheduled, the delivery rate was increased to 55 MCFD between 2:00 and 3:30. For an hour thereafter, the delivery rate was erratic, with surges well above the rated capacity, a condition under which there was liklihood of escape of hydrogen sulfide to the atmosphere, most likely from one of the Girbotol units. After the plant was shut down, we inspected the nozzles and supply lines, the flare and found them partly filled with a sooty residue with traces of monoethanolamine, which suggests that under surge conditions occurring that morning, the amine solution again overflowed and impaired combustible gas delivery to the flare, which also could have allowed some of the hydrogen sulfide enriched gas to escape.

The weather was cooler than usual and the winds weak. We invited an epidemiologist to study the episode, so that together we can arrive at some recommendations for preventing future technological disasters.

III. THE POLLUTANT AND ITS EFFECTS

Hydrogen sulfide is a heavy gas (1.54 g/ℓ, compared to water at 1.00 g/ℓ) and relatively soluble in water. At low concentrations its characteristic, rotten-egg odor is very familiar, but at higher concentrations (1 to 10 ppm) the awareness of odor is overwhelmed and its irritating properties for the conjintiva and respiratory tract begin to be noticed, with cough and respiratory irritation and nausea and vomiting dominating the picture up to about 1000 ppm (0.1%). Between about 1000 and 2000 ppm neurological effects begin, with effects interfering with respiratory reflexes, and resulting respiratory paralysis being a mechanism for fatal outcome in acute poisoning. The material is relatively rapidly oxidized in the body, and survivors of high exposures usually have no sequelae. In high enough exposures, it can be as rapidly fatal as hydrogen cyanide.

Exposures occur, in addition to the petroleum refining industries, as a result of reactions of salt water and sewage or garbage, and in cleaning cess pits. Natural mineral and geothermal springs contain relatively small amounts of hydrogen sulfide, which have occasionally caused fatal poisoning. Usually severe or fatal exposures affect a small number of persons, and no other community disasters such as this were known.

Because of its rapid disappearance in the body, no specific therapy is indicated other than providing clean air and treatment of the pulmonary irritation or edema. Artificial respiration

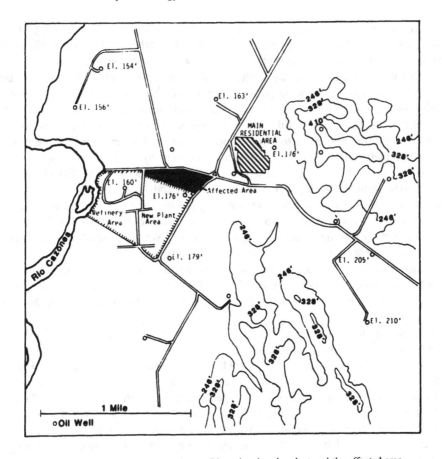

FIGURE 1. Topographic map of Poza Rica, showing the plant and the affected area.

is lifesaving if promptly used, but in a situation where high concentrations exist, the person who tries to help or resuscitate a victim is himself at great risk. In too many industrial exposures, would-be rescuers of victims have themselves become victims, so training in the use of gas masks or, better, air-supplied breathing masks is essential in industrial safety programs where hydrogen sulfide risks are to be expected. There is some question as to whether high exposures may lead to irreversible changes in the nervous system among those who can be resuscitated.

IV. THE COMMUNITY

Figure 1 shows the location of the plant and the area where the population was affected. At the time of the disaster there was an atmospheric temperature inversion with winds from the southwest at 4 knots (4.6 mi/hr). Heavy fog patches were reported. Under these inversion conditions, emissions of steam and smoke tended to move horizontally from the plant to the affected residential area. Most of the dwellings were built of split bamboo, some with thatched roofs, and others with galvanized sheet metal. Many had dirt floors, but some had foundations which elevated the floor about 2 ft. Some of the dwellings in which fatalities occurred were within 200 ft from the plant or from the sulfur recovery unit, located near the boundary between the plant and the affected area.

V. THE EPIDEMIOLOGICAL DATA

Three types of data were analyzed. They are descriptive information about when, where, and how selected people were stricken; a description of the fatal cases by age, sex, time of death and in one case an autopsy, and the review of hospitalization data for 47 out of the 320 people who were hospitalized.

Twenty-two persons died. The epidemiological team arrived and worked for two days, December 7 and 8, about 2 weeks after the episode on the morning of November 24.

The episode started at 4:50 am and by 5:10 the sulfur recovery unit was shut down, and army and police were notified, and rescuers without any protective equipment were able to work in the area, only experiencing nausea. The episode lasted 20 min.

VI. CASE REPORTS

In a house about 200 ft from the sulfur removal unit, a 37-year-old man awoke and noticed that his 13-year-old daughter was ill. She got up and attempted to come to him, but fell to the floor. His 2-year-old son also fell to the floor when trying to come to his parents' bed. The man noticed burning of the eyes and a severe headache. He was able to get up, wet some blankets, which he put over the heads of the children; with his 24-year-old wife, he was able to carry the children out of danger, and went by taxi to a nearby community. All recovered completely by the next day.

In another house about 400 ft from the sulfur removal unit, a 45-year-old father awoke, saw that something was wrong, carried his 4-month-old baby out of the area, but both were hospitalized. His wife, 37, and three older children did not leave and died in the house.

In a house about 600 ft from the plant, a mother awoke and found that she was partly paralyzed. She called to her daughters aged 17 and 25 to get help, but as they opened the door to leave they fell unconscious, and were pronounced dead on arrival at the hospital. The mother and three children aged 5 to 12 remained in the house and though hospitalized, recovered.

A 26-year-old woman living 200 ft from the plant was brought to the hospital unconscious, remained in a coma for 7 days with tonic and clonic convulsions of upper and lower extremities, evidence of kidney failure, and died on the 7th day. Two of her children also died, but her husband and three children survived.

An 11-year-old girl awoke, felt ill and went into the bed of her parents. All three were found unconscious and on arrival at the hospital the girl was dead but the mother aged 24, and the father aged 32, recovered.

Of the fatal cases, 9 were dead on arrival, 4 died within 2 hr, 4 more by 6 hr, and 1 each 24 hr, 2 days, 5, 7, and 9 days after admission. Twelve females and 10 males died. Nine were children under 13 years, 10 were between 14 and 35 years, and 3 were between 36 and 50 years old. The autopsy of a 22-year-old woman showed edema and congestion of the lungs, many small hemorrhages in the brain with congestion. The spleen, kidney, and heart were also edematous.

The frequency of symptoms and signs among the 47 hospitalized patients, whose records were reviewed is shown in Table 1. Of the 320 persons admitted to the hospital, 170 were discharged within 3 hr, and additional 90 within 7 hr. While not all patients showed effects on the central nervous system, those who did had a high case-fatality rate.

Four survivors had sequelae, two with acoustic neuritis, one had dysarthria (difficulty in forming words), and the fourth, an epileptic has had aggravation of his disease.

The oldest of the 47 was an 84-year-old woman, who had been unconscious and recovered completely.

Several possibly useful types of information are not available as a result of this epidemiological study, particularly estimates of the populations living at various distances from

Table 1
SELECTED SYMPTOMS AND FINDINGS OF 47
PERSONS HOSPITALIZED AS A RESULT OF THE
HYDROGEN SULFIDE EXPOSURES AT POZA RICA

Age	Sex	No.	Loss of sense of smell	Uncon-scious	Severe headache	Vomit-ing	Nausea	Cough or dyspnea	Pul-monary edema
<10		5	5	1	2	1	1	4	—
11—20	M	3	3	1	1	1	1	1	1
	F	2	2	1	2	1	—	1	—
21—30	M	5ª	5	2	1	1	1	1	1
	F	7	7	4	7	2	3	6	—
31 +		7	7	5	4	2	1	7	1
Age unknown									
"Minor"		2	2	1	1	—	1	1	1
"Adult"		8ª	8	4	6	3	3	4	1
No entry		7	7	5	4	3	4	2	2
Total		46	46	25	28	14	15	27	10

ª   One of these died in the hospital in each group.

From McCabe, L. C. and Clayton, G. C., *AMA Arch. Ind. Hyg. Occup. Med.*, 6, 199, 1952.

the presumed source of the gas. The entire area had a population of 22,000, but it is obvious that only those living close to the plant were at high risk. Yet we are not told the population of the "affected area" in Figure 1. The study team had but a short time to prepare for the study and according to the report may have spent as little as 2 days in the community. Furthermore, if populations at risk were better known, would that have made possible some more useful conclusions?-probably not. There was no uncertainty as to the nature of the pollutant nor its source. The report states: "The victims were affected in a geographical area in direct proximity to the effluent stack of the Girbotol unit."*

The report makes three recommendations: that hydrogen sulfide monitors should be located at strategic areas within the plant and be capable of giving an alarm; that a strong industrial health and safety program be developed; and that an air pollution study be done in the region as there seemed to be evidence of possible hazards from both hydrogen sulfide and sulfur dioxide.

These recommendations seem to avoid two issues that might be critical to prevention of future disasters; the distance of dwelling units from possible sources of highly toxic pollutants, and the procedures for prevention of malfunction of equipment and for correction of possible design errors, once some malfunction had been detected. Of course, the publication of a scientific article on such a disaster can be considered relevant to possible litigation, and the authors may have decided that public disclosure of recommendations which implied some level of negligence would be inadvisable.

*   From McCabe, L. C. and Clayton, G. D., Air pollution by hydrogen sulfide in Poza Rica, Mexico: an evaluation of the incident of Nov. 24, 1950, *AMA Arch. Ind. Hyg. Occup. Med.*, 6, 199, 1952.

VII. POLICY IMPLICATIONS OF THE STUDY

This report of a disaster in Poza Rica, Mexico on November 24, 1950, published in September 1952, over 30 years ago, is unique in that it starkly documents the potential hazards of a common gas, hydrogen sulfide, the risks of residents in the immediate proximity of industrial processing plants, and the inadequate attention to correcting a problem identified in breaking in a new system. If such a refinery disaster occurred in the 1980s it is most unlikely that it would be accurately reported in a scientific or technical journal. On the other hand, there is no evidence that such a disaster has recurred in spite of the fact that the opportunities have greatly increased with the expansion of refining facilities. Exposures to hydrogen sulfide continue to claim lives in industry, and odor problems from lower concentrations continue to be problems for some communities, some such effects are discussed in Chapter 11. Whether the experience gained in this disaster is partly responsible is almost unanswerable. We also do not know of additional research which this study has stimulated.

The importance of this community study lies in its prototypical nature. A new technology was introduced in a developing region; housing was very close to potential sources of toxic materials; the malfunction responsible occurred at night, and the dissemination of the heavy gas was more harmful because of low winds and inversion conditions. We can assume that those exposed would have had a higher dose if they were physically active, since with such activity the respiratory rate increases, and the more air taken in the greater the exposure dose.

The case reports mention blankets and wet cloths as possible protective measures, but the data do not permit any adequate evaluation.

A final lesson can be mentioned; those who live near industrial process facilities need to know what materials are being processed, what their hazards may be, and what can be done to protect themselves. The recommendation for stronger industrial hygiene and safety programs could fulfill this requirement if it is to be assumed that those living in proximity to such plants are employed at the plants. The community study doesn't provide us any information that will confirm or refute that the residents in the affected area knew the hazards of hydrogen sulfide, nor does it tell us that if they were better informed they would have been better protected. In the case of this agent, knowledge of specific medical treatment was of little importance, but for other potential industrial spills, this may be the critical piece of information.

Of the acute hazards of chemical exposure that have been reported on the news media recently, one gets the impression that more have been from spills associated with transport or storage, such as tanks, tank cars, tank trucks, or pipelines. Such hazards involve a much more diffuse community and different preventive strategies. Police, highway patrolmen, and firemen become the most critical group for community protection as a result of chemical transport accidents.

The tragedy in Bhopal, India has additional lessons concerning testing of safety procedures to teach us.

Chapter 6

THREE LOS ANGELES HEAT WAVES

John R. Goldsmith

TABLE OF CONTENTS

I. THE PROBLEM

That extremes of heat and cold can result in excess mortality is widely known. Yet surprisingly, few studies have attempted to put the problems in a quantitative way or derived information which can be useful for prevention of the excess mortality. Since such extremes of weather are bound to recur, there is abundant reason to carry out such epidemiological studies. One of the three heat waves to form the basis of this analysis occurred in 1955, and was preceded and followed by severe photochemical air pollution. It remained a hotly debated question as to how much the exposure to pollution contributed to the excess mortality. By comparing the excess during the 1955 heat wave with that found for a heat wave in 1939, before photochemical pollution was much of a problem, we hoped to help answer this question (see Chapter 1, Figures 1 and 2).

Previous studies had made some important contributions to our understanding of the problem.[1,2] Heat waves when they occurred tended to affect broad geographic areas, and could occur anywhere; however in different locations, populations had different levels of adaptation, based on the prevailing environmental conditions, modified in many cases by housing and cultural practices. Different locations tended to have heat waves with different levels of humidity, and it was understood that at a given level of high temperature, the higher the humidity the greater the risk of collapse or death, since heat loss by sweating was more difficult at high humidity. In general, heat waves meant several days in succession when the maximal temperature exceeded the mean body temperature of 98.6°F (37°C).

In heat waves, when excess mortality occurred, very few of the excess deaths were attributed to heat, heat stroke, or heat exhaustion by the physician who was required to state the cause of death. Therefore, all causes of death must be studied together in order to adequately reflect the risk.

The proportion by which mortality increases rises in a nonlinear way both with age and temperature.

Maximal temperature (as for example compared to mean or range) is the single meteorological value with the greatest predictive value, and its relationship is predictive of the mortality for the following day. This one-day-lag phenomenon may either reflect the dominance of relatively slow mechanisms or the fact that we are studying successive days in each heat wave and effects may cumulate; both mechanisms may be in operation.

II. THE COMMUNITY SETTING

Los Angeles is a large metropolitan complex situated on a coastal plain, bounded on the north and south, to some extent by relatively high and steep mountains. During the period of this study it grew very rapidly in population, and with the development of its freeway system, developed a dominant pattern of urban transport based on the private passenger automobile. Between 1945 and 1950, increasingly severe air pollution, called "smog" was observed which had the property of oxidizing materials or chemical reagents. The oxidizing ingredients were showed to be due to the action of sunlight on a mixture of hydrocarbon vapors and oxides of nitrogen; the most characteristic product was ozone (O_3) known to be a strong oxidant. The severity of the pollution by photochemical products has now come to be measured by the levels of ozone. The accumulation of high amounts of the photochemical ingredients, hydrocarbon vapors, and oxides of nitrogen, is greater during periods of low winds or atmospheric stagnation.

The prevailing summer meteorological condition is a gentle offshore wind, with mean maximal temperatures in the range of 80 to 85°F, with an offshore high pressure area.

Occasionally, the high pressure area moves inland and then the winds reverse and come from the southwest, and are hot and dry, with maximal temperatures over 100°F and minima above 80°F. Under these so-called "Santa Anna" conditions, the humidity is very low, and winds may be strong with minimal air pollution.

Such conditions usually occur in the month of September, and all of the episodes reported here are in that month.

In a large community such as Los Angeles, there are many persons in frail health, for whom even an environmental exposure which would be considered only a mild one by a person in good health may lead to a destabilizing process with a fatal outcome. Such unusually sensitive groups include both the very old and very ill, as well as persons with severe acute illness or injury from which recovery is problematic. In thinking of increases in deaths due to heat waves or to smog attacks, it is both realistic to think that a high fraction will be taken from among those with precarious health — and it is also erroneous to think that the health toll is limited to such persons. Weather or pollution which is sufficient to increase the chances of mortality of the severely ill may also disable or impair those with more robust health. It is therefore realistic to think of a community like Los Angeles as including people with a spectrum of health statuses; when heat waves or smog attacks occur, mortality will be greatest among those with the poorest health status, but health impairment may extend to those with better health, whose mortality rate may be only very slightly increased. It further follows that measures available to prevent mortality among those most susceptible may also be measures which sustain the health and vigor of a much larger segment of the population. Such reasoning gives added importance to studies of episodes of increased mortality from environmental causes.

III. THE 1955 HEAT AND POLLUTION EPISODE

Figure 1 shows the dramatic increase in mortality occurring in early September 1955, preceded and followed by several smog alert days, days when the maximum hourly ozone level was in excess of 0.5 ppm. The temperature maxima reached the following values (in °F) on successive days: 101, 110, 108, 103, 101, 102, and 100. As can be seen in the figure, several days together with ozone alerts did not produce detectable deviation in mortality, but the deaths increased sharply when the temperature rose and dropped just as sharply when the temperature declined despite the occurrence of smog alerts. Neither did isolated ozone alerts have any detectable increase in mortality. The data are for deaths of persons over 65 years of age, which include many persons in frail health.

During this same period, the California Department of Public Health was operating an epidemiological monitoring of nursing home deaths, in order to detect if there was any increase in mortality of the unusually vulnerable population associated with increased air pollution. All nursing homes with 25 or more beds reported to the Department each week the number of deaths and transfer to hospitals of their residents. The assumption was that if some increase occurred in such deaths or transfers, it might be a sort of early warning of possible harm to more health persons. Figure 2 shows the same episode, with a slightly different scale for the maximal temperature, but instead of deaths of persons over 65 it shows the deaths of persons who had been nursing home residents. The relative sensitivity of these two systems can be judged roughly by the proportional rise during the heat wave from about 60 to 317 in the case of all persons over 65 years of age, a 5.3-fold increase, compared to a rise from about 6 to 48, an eightfold increase for the nursing home population. However the nursing home population shows no apparent impact of the smog alerts either, whether associated with the heat wave or at other times. The nursing home population monitored averaged about 4000 persons.

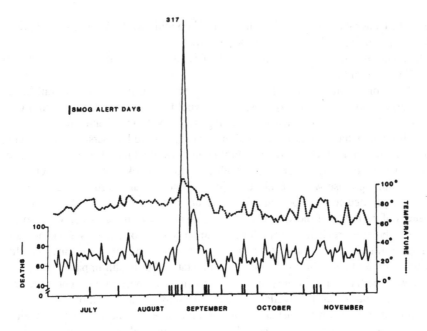

FIGURE 1.   Association of mortality of persons over 65 years of age and temperature during the heat wave of 1955. On the days marked "smog", alerts occurred and the ozone levels exceeded 0.5 ppm. (This figure is identical to Chapter 1, Figure 2.)

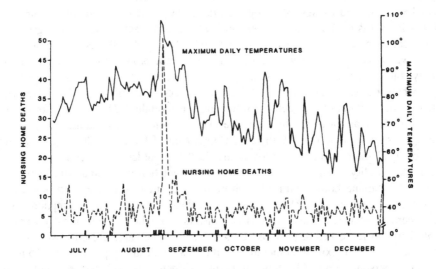

FIGURE 2.   Association of mortality of persons resident in nursing homes in Los Angeles and heat during the heat wave of 1955. "Smog" alerts occurred during the days indicated. Note that the temperature scale is different for this figure, but the same temperatures are recorded. (This figure is identical to Chapter 1, Figure 1.)

IV. HEAT WAVE MORTALITY FOR 1939 AND 1963

The 1939 heat wave was in late September with successive days of maximal temperature (in °F) of 100, 103, 104, 107, 106, 103, and 101. It was as the 1955 episode, of 7 days duration, with the same mean temperature of 103.6, but a slightly lower range. The 1963 episode, also in late September was shorter, but with greater mean temperatures, viz; 107, 109, 106, and 102, with a mean of 106°F. Except for 1947, a "normal" year with maximal temperature of 96 during the period, these events seem to occur each 8 years.

V. ANALYTICAL APPROACH

Copies of death certificates for each death occurring among persons 50 years and over for each of these 4 years including 1947 were obtained for residents of Los Angeles and Orange Counties (Orange County is the county just south of Los Angeles, also mostly a coastal plain, and sharing the same meteorological patterns) for the period of August 21 through October 10. These dates were selected in order to obtain data for a "normal" period preceding and following each episode. Data for total mortality by 5-year age groups and daily maximal temperature were tabulated.

Since the population changed greatly in numbers and in age structure during the period, absolute mortality rates, (deaths per 100,000 persons) would have required extrapolation of census data, which could lead to erroneous estimates. Instead, a mortality ratio of the observed to expected was calculated; the expected mortality was computed by averaging the daily mortality for the age group in question for the 42-day period excluding 9 days each year beginning with the first day of the hot spell. For 1947, an arbitrarily chosen 9-day period was excluded, since there were no days with maximal temperatures in excess of 100°F.

VI. RESULTS

Two obvious facts emerged from the resulting proportional mortality ratios; the ratios increased with both maximal temperature and increasing age, and there was a 1-day lag between the peak of temperature and the peak of mortality in each age group. Some exceptions to these generalizations were observed and will be noted. In the numerical analysis, we refer to the Age-Temperature Specific Mortality Ratio (ATSMR) as a percent of expected.

In 1939, the expected daily mortality for all ages over 50 was 55.9 persons. The initial peak temperature of 103°F was followed a day later by a mortality of 193% of expected. The later peak was followed a day later by a mortality of 271% of expected. With some irregularities, the ratios increase for the duration of the episode by age group; for the 50 to 54-year-old, the ratio is 245%, and for those over 85, 570% of expected.

The total excess mortality for the 9-day period was 546 deaths or about 61/day. The hot spell was broken by a most extraordinary tropical storm with $5\frac{1}{2}$ in. of rain and winds of 47 mph, which was reported to have killed 45 people, about 10% of those killed by the heat wave.

In 1947 there was no excessive heat and no evidence of any deviation from expected mortality. The maximal deviation was 132% of expected, which is well within the 99% confidence limits, based on the Poisson distributional assumptions. The expected daily mortality was 75.

For 1955, the expected daily mortality was 86.4 deaths per day. The initial peak of 110°F was followed one day later by 385 deaths, or 445% of expected. The later rise of 100 to 102 was followed one day later by a rise to 160% of expected. Maximal mortality for the 50 to 54 age group was 307% of expected and for the over 85-year-old group 810% of

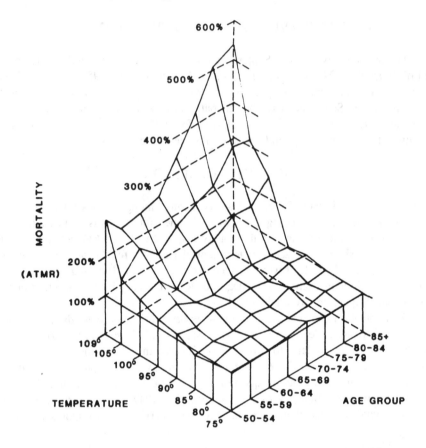

FIGURE 3. Observed mortality by age and temperature (atmr) of Los Angeles and Orange Counties combined during the heat waves of 1939 and 1955 (data combined). (From Oechsli, F. W. and Buechley, R. W., *Environ. Res.*, 3, 277, 1970. With permission.)

expected. The excess for the 9 day period was 946 deaths or about 105 per day. The patterns were generally similar to 1939 for ATSMR.

For 1963, there appeared to be some moderation in the expected impact of high temperature. First the mortality peak of 172% was reached 3 days after the peak in temperatures of 109°F. Despite the greater mean temperatures, the mortality ratios did not reach such a high level as in the two earlier episodes; the increases with age were not as steep, being 140% of expected in 50 to 54-year-olds and 257% among those over 85. The total excess was about 580 deaths or about 50 per day.

Figure 3 and Figure 4 show the results for 1939 and 1955 combined in the form of a three dimensional diagram as observed in Figure 3 and as computed by an empirical prediction equation based on an exponential function:

$$\text{ATSMR} = 98.806 + \exp(-15.23 + 0.0385 \text{ age} + 0.1655 \text{ temperature}) \quad (1)$$

It is clear from the comparison of Figure 4 to Figure 3 that the equation fits the data quite well. Figure 5 shows the data for the 1963 heat wave, and the differences are striking. Fitting the same type of exponential function to these data gives:

$$\text{ATSMR} = 90.711 + \exp(-6.557 + 0.0338 \text{ age} + 0.0770 \text{ temperature}) \quad (2)$$

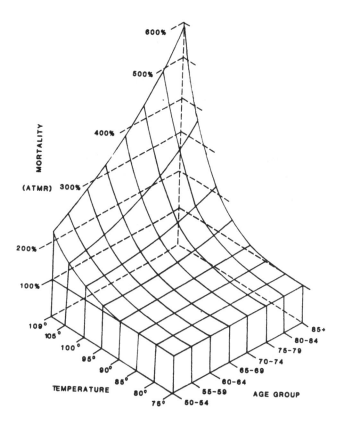

FIGURE 4. Fitted mortality by age and temperature according to a regression equation for the data shown in Figure 3. (From Oechsli, F. W. and Buechley, R. W., *Environ. Res.*, 3, 277, 1970. With permission.)

The age coefficients are similar, but in 1963 (Equation 2) the temperature coefficient is much smaller.

VII. INTERPRETATION OF THESE FINDINGS

First, the magnitude of the excess mortality in each of these three episodes is very great. Excluding infectious disease epidemics, each of the heat waves exceeds all other natural disasters recorded in the state of California in the number of excess deaths. For example, the great San Francisco earthquake and fire of 1906 resulted in 452 deaths.

The relatively modest increase in mortality in the 1963 episode is noteworthy, and was quite unexpected. Kutschenreuter in 1960[3] proposed the most likely explanation, the introduction of air conditioning. Prior to 1959, the peaks in the use of electricity in Los Angeles occurred in the winter, but since then the peaks have been in the summer. The most probable explanation is the shift in using electricity for heating units to its use for air conditioning. During the 1963 hot spell an all-time maximum in use of electricity occurred. Similarly, during heat waves in New York and St. Louis in 1966, electricity use threatened to overload the distributional system.

If this hypothesis is true, the use of air conditioning could have been responsible for the saving of about 800 lives, which is the number of excess deaths which would have occurred if the conditions of Equation 1 rather than Equation 2 had prevailed.

This is consistent with the demonstration from the pre-air conditioning period that excess deaths among elderly persons in nursing homes could be prevented by monitoring their

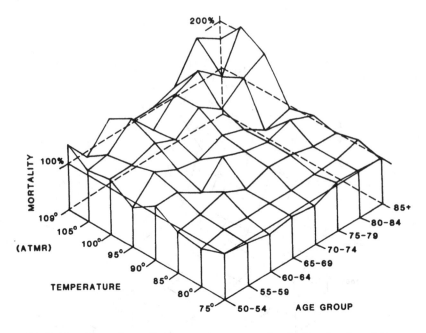

FIGURE 5.　Observed mortality by age and temperature (atmr) of Los Angeles and Orange Counties (1963 heat wave). (From Oechsli, F. W. and Buechley, R. W., *Environ. Res.*, 3, 277, 1970. With permission.)

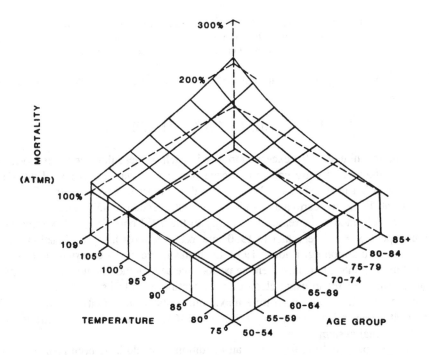

FIGURE 6.　Fitted mortality by age and temperature according to a regression equation to the data in Figure 5. (From Oechsli, F. W. and Buechley, R. W., *Environ. Res.*, 3, 277, 1970. With permission.)

ability to sweat or perspire. Those who were not able to perspire when environmental temperature exceeded normal body temperature were likely to have an increase in body temperature and a high risk of heat stroke or other cause of mortality. Older persons unable to perspire, could be protected from heat risks by using wet cloths and monitored to see that their temperature did not rise. Thus, in the absence of air conditioning, good nursing care based on physiological principles can also prevent heat wave mortality and morbidity. Impaired circulation or kidney function may be one factor in the poor adaptation of older persons to heat; in addition diuretic drugs or other modes of treatment such as low salt diets can impair the adaptability of such persons.

REFERENCES

1. **Gover, M.,** Mortality during periods of excessive temperature, *Public Health Rep.*, 53, 122, 1938.
2. **Oechsli, F. W. and Buechley, R. W.,** Excess mortality associated with three Los Angeles September hot spells, *Environ. Res.*, 3, 277, 1970.
3. **Kutschenreuter, P. H.,** Weather does affect mortality, *ASHRAE J.*, p. 239, 1950.

Chapter 7

PLANNING, ANALYSIS, AND PREPARATION CAN SAVE LIVES

John R. Goldsmith

Chemical spills are not all that rare, and most of them are not lethal; what was so unusual about the Poza Rica Disaster?

The first thing unusual about the Poza Rica disaster is that we have one scientific report. When PCBs (polychlorinated biphenyls) contaminated dairy feed in Michigan and this highly persistent material entered the human food chain, there were many reports, and the damage is still being assessed. In Seveso, Italy, a pesticide plant's reactor vessel burst, contaminating the local countryside with dioxin and spreading fear of birth defects among the parents and expected parents. This is still the subject of articles and reports.

Faulty waste disposal led to commonly heard phrases, "Love Canal", Kepone dumping in the James River in Virginia, Stringfellow acid pits in California. Why did we not use one of these to illustrate the risks posed by the chemical industry for the community?

The choice of the Poza Rica story was because the disastrous impact was not due to the spill itself as much as to the *close proximity of homes to a major chemical plant*. This risk is avoidable, and it requires planning to avoid it. It often happens that housing is built up in the vicinity of chemical plants which were located correctly at a considerable distance from populations at risk. In such a case it is the land developers who are to be criticized and the planning authorities who bear most of the blame. In other cases, the choice of the plant location is faulty (for example the pulp mill upwind of Eureka, Chapter 1). Dilution is usually a power function of distance from the source: Concentration at distance $D = E \times D^N$, where E is the emissions, and N is between 1.5 and 4.

In a community with a refinery and/or chemical plant, it is important for the residents to know what the materials are that they may be exposed to and what measures they may take to protect themselves. This principle of "the right to know" applies especially strongly to workers. There is little evidence that the employees of the Poza Rica refinery knew what protection was needed for exposures to hydrogen sulfide. In most cities, such knowledge is especially important for fire departments, since either the firemen may have to fight fires in such plants or they may have to provide rescue or resuscitation for persons exposed.

What is the situation in your community? Are there any chemical plants? What material do they handle which might be a hazard for the community? What information does your local fire department have concerning the hazards? What do you think should be the role of the health department?

One of the major debates in connection with dams, flood control projects, and nuclear plants, is how much investment is justified in order to avoid hazardous breakdowns. What do you think is the proper basis for requiring industries to invest in pollution control and community protective planning and engineering? Do you think that all such accidents can be prevented? Why?

Should an industrial plant in an underdeveloped country have the same stringency for worker protection and community protection as it might be required to have in the U.S.? If not, should it have greater or less?

Preventability of health damage from heat and cold waves — The analysis of three Los Angeles heat waves provides a convincing picture of two facts; older people are more at risk during heat waves and something happened between 1955 and 1963 to make the impact less. Our best guess is that air conditioning is largely responsible. If air conditioning is capable of reducing the community risk from heat waves, what other actions should follow? Not everybody all the time can stay in air conditioned rooms or buildings, but we can identify in advance the people with heart and kidney disease, and even if they cannot always be in air conditioned places during heat waves, it is possible to take other preventive measures. These include decreased exposure in the sun, careful adjustment of salt and water intake, and monitoring to see that the mechanisms that regulate body temperature are working. An example of this is the program adopted in a New York City nursing home to regularly check each resident to make sure that they were sweating during hot spells. (This program was adopted prior to the widespread use of air conditioning.) The circulatory collapse which is often the most serious event in heat stroke is often preceded by a sharp reduction or cessation in the secretion of urine. To prevent this it is advisable to make sure that fluid intake is large enough so that frequency and volume of urination don't decrease during a heat wave. While the analysis and this discussion so far have emphasized deaths during heat waves, these same principles apply to anyone maintaining physical vigor during a heat wave. It is virtually certain that during a heat wave with excess deaths, there is a great deal of excess illness as well.

Look up in newspapers or magazines for accounts of a recent heat wave in your community. Was there an excess of deaths, compared to previous years at the same time (or subsequent years if the heat wave was several years ago)? Are the homes for elderly persons in your community air conditioned? Are there special steps that they take during heat waves? What are they? Do such steps require advance preparation?

When excess deaths do occur during heat waves, do they occur on the first day? If there seems to be a delay in the excess mortality, what does this suggest as to the time available for helping the potential victims to adapt?

Most heat waves are forecast by the weather service. Should such forecasts provide advice to elderly and ill persons to take preventive measures? If you answered yes, what measures, at what temperature level, and who should offer the advice?

Although the mechanisms and time of year are different, most of the same principles apply to exposures to cold.

Elderly persons are usually the victims of cold and heat waves. Those with adequate heating in their homes are protected, but for many urban and rural poor, the cost of heat is high and sometimes they are not able to provide heat for their apartments or rooms.

A drop in body temperature is the danger sign, and rewarming is the treatment. Most thermometers available were made for the purpose of finding out if a person has a fever. A low reading thermometer is just as simple as a fever thermometer, and its use can provide the needed warning.

Babies can suffer a serious or fatal hypothermia during cold periods, and the reports of this occurrence are more abundant from Mediterranean, subtropical, and Third World countries than in Europe and North America. A possible reason for this is that household heating is rarely needed for adults in such countries, and the vulnerability of small infants to cold is not appreciated by the parents. The so-called "cold injury syndrome" is more frequently seen during the cold season in babies of relatively low birth weight, and of course during cold weather.

Home visits by Public Health Nurses equipped with low reading thermometers are the key to prevention.

Can you find reports of excess deaths during the most recent cold snap in your community? If so at what ages and for what causes? What is the proper treatment for hypothermia? (That is defined as body temperature below 35°C.)

Is there any program for prevention of deaths or injury from hypothermia in your community? If not, what elements should it contain?

Part C

Chapter 8

RESPIRATORY PROBLEMS AMONG KAWASAKI SCHOOL CHILDREN

Toshio Toyama

TABLE OF CONTENTS

I. INTRODUCTION

This chapter describes a now classic study of respiratory effects of air pollution. The epidemiological design consisted of repeated cross-sectional observations (prospective study) on school children from two relatively localized areas: one exposed to extremely heavy pollution, the other clean. Observations of pulmonary functions of the two groups were compared to determine how these functions varied with the magnitude of exposure to a causative agent (exposed-referent study).

Unlike reports of the Great London Smog (1952) which detailed mainly mortality due to a chemically reducing type of air pollution, this study of Kawasaki, sought to observe, in a short term, early stages of pulmonary disease.[1] Although both cities suffered from pollution due to coal, the Kawasaki pollution differed in its lack of a significant contribution by household smoke. This study also sought to offer evidence in support of measures for public health protection and environmental control. An overview of the study was published elsewhere in 1964.[2,3]

II. BACKGROUND

A. Kawasaki City

Kawasaki, located between Tokyo and Yokohama, is a city of heavy industries. Although most of the city was devastated during the war, the large industrial plants, coal-fired power stations, iron works, and shipyards on the east side of the city facing Tokyo Bay survived, and resumed operations about 5 years after the end of the war. Due to lack of emission abatement facilities and rapid industrial growth, air pollution, mostly of a reducing type (sulfur oxide and suspended particulate matter), soon reached an unfavorable level and became a matter of local public health concern in 1955.

One meteorological characteristic of the city is a regular seasonal variation of wind direction: in spring and early summer the winds are SSE from the Bay and in winter NNW. Therefore, city inhabitants, except those in a far western hilly area of the city, had been regularly exposed to heavy pollution in the spring and early summer. Very high population density, 7500 people per square kilometer, was another difficult factor in the health problem. For reference, the site of the "Tokyo-Yokohama Asthma" (1954), and epidemic of acute environmental pulmonary ventilatory disorder among American military personnel and their families, was about 30 km west of Kawasaki.[4]

B. Air Pollution and the Public School in the Study Area

In January 1957 Keio University Medical School began measuring two indexes of air pollution, monthly sulfation rate* by lead peroxide method and monthly dustfall (total suspended particulate matter) by British standard deposit gauge.[5] These were measured on the roof of a three-storied primary school building, denoted as H-School (Figure 1), in a heavily polluted area located about 1 km northeast of the heavy industrial complex. A doctor at the city hospital served as an intermediary to enlist the headmaster's cooperation and gain his acceptance of the project. The measurements were continued monthly for 10 years until 1967, which was undertaken by medical students as part of preventive medicine laboratory work under the guidance of the author. The results of the monthly sulfation rate and dustfall for 10 years are shown in Figure 2.[6] Later in 1962, the national and city air monitoring networks were inaugurated by the Smoke Control Law.

---

* This is a passive monitoring system in which the sulfur oxides react with a paste of lead peroxide to produce lead sulfate, which after a month's exposure can be analyzed by wet chemical methods.

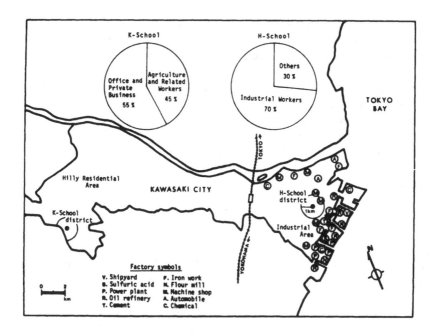

FIGURE 1. Map of Kawasaki city, location of H-School in industrial area and K-School in residential area, and circle graphs of parent's occupation of the school children.

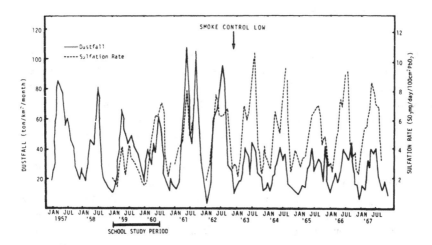

FIGURE 2. Ten-year variation in monthly air pollution levels at H-School in Kawasaki industrial area.

It should be noted that dustfall showed a clear seasonal variation every year from 1957 to 1962 fluctuating from peaks of more than 80 tons/km²/month in early summer in 1957 and 1958 down to 10 to 20 tons/km²/month in winter. From 1962 on, the dustfall dropped sharply due to conversion from coal to oil followed by installation of particulate abatement facilities. As a consequence, however, gaseous sulfation rate increased and followed a seasonal variation. Peak values of the dustfall were extremely high compared with those in other cities: 30 tons/km²/month in New York (1956), 14.4 tons/km²/month in Los Angeles (1948) and 39.0 tons/km²/month in London (1954).[7]

H-School was a public primary school for children who lived within a "school district", an area within a radius of about 600 meters. Since more than 80% of the students' homes were in this district, and the school in its center, the students were exposed to constant and uniform air pollution throughout the day either at home or at school. According to the headmaster, the children complained daily about books becoming soiled, a thin layer of dustfall accumulating on desks, occasional cough and sputum, and eye and throat irritation. These conditions were observed even when the window of the classroom was closed, and since air conditioners were not installed in the school their complaints grew more frequent in early summer. Our proposed epidemiological study of student health was accepted after long consultations over the unusual situation with the headmaster, thus ensuring his cooperation and that of his students. We were able to start collecting the necessary data in 1959, continuing for 18 months without any constraint from the outside and with our own funding from the university.

III. OBJECTIVES AND STUDY DESIGN

A. Objectives

The aim of the study was to explore the effects of air pollution, sulfur oxides and dustfall, on the pulmonary ventilatory capacity of the H-School children in the polluted area by comparing Wright peak flow rate and other pulmonary function parameters of H-School children with those for reasonably matched K-School children in a less-polluted area in Kawasaki, the measurements to be repeated as frequently as possible and simultaneously over a relatively short period of time, at least more than 1 year. And, the data collected were to be analyzed to determine if there was significant quantifiable variation of ventilatory capacity between the exposed and referent cohorts in every specified month in proportion to the magnitude of air pollution.

B. Subjects

In order to match exposed and referent subjects with respect to age, sex, and stature, all students of both sexes in the fifth grade, age 10 to 11 years in June 1959, were selected in two public primary schools: H-School in the polluted area and K-School in a less-polluted area in Kawasaki City. The number of subjects at the beginning of the study in February 1959 was 48 males and 68 females for H-School, and 48 males and 45 females for K-School. In February 1959, the actual grade of the subject was the fourth in both schools, since in Japan the new school year begins in April and ends in March. The details of height, weight, and number of students classified by age, sex, and school grade are given in the tables of Section V. The reasons for selection from the higher grades at the beginning of the study were (1) that a 5th grader could be followed up during the 6th grade after April 1960, since the study had to continue more than 1 year and (2) that better cooperation was expected from higher grade students in the pulmonary function test which usually required coordinated effort, thus ensuring more reliable data. All the subjects were nonsmokers and were served the same type of school lunch. Eighty percent of their residences were within the relatively limited school districts where they were exposed to different levels (high in H-School, low in K-School) of air pollution constantly, whereas their lifestyles and socioeconomic conditions were almost the same, except for some difference in parents' occupation. The majority of parents (70%) were factory employees in the H-School area, and white collar or agricultural workers in the K-School area as shown in the graphs in Figure 1.

IV. DATA COLLECTION

A. Air Pollution Measurement

Monthly sulfation rate by lead-peroxide cylinder and dustfall by standard deposit gauge were measured continuously at H-School during the study period. In K-School only dustfall was measured in the same period for control. As shown in Table 1, in H-School the dustfall increased conspicuously in the spring and early summer months up to 70 tons/km²/month in April 1959, then dropped sharply in winter to 15.7 tons/km²/month in January 1960, and rose again to 61 tons/km²/month in August. The sulfation rate also showed the same trends. It was clear that the remarkable seasonal shifts were due to a dominant seasonal wind direction as characterized by topographic conditions as shown in Figure 1 and Table 1. This seasonal variation followed the same pattern recorded in the years preceding the study period (Figure 2). On the contrary, in the K-School area the dustfall fluctuated around 10 tons/km²/month throughout the year without appreciable seasonal variation. Chemical composition of the dustfall in H-School was mainly coarse coal fly ash with a median diameter of 2.0 μm.[2] Although both sulfation rate and dustfall are crude indicators for classic type air pollution, they were thought appropriate to use in a relatively small community as collective environmental indexes.

B. Pulmonary Functions

As the dependent variables of the effects of exposure to air pollution, peak flow rate by Wright's peak flow meter (PFR or expiratory flow rate), vital capacity (VC, unforced expiratory), and 0.5-sec forced expiratory volume ($FEV_{0.5}$) by a water-filled spirometer were measured in children of the two school, except $FEV_{0.5}$ in K-School.[8] The repeated function tests were employed regularly every few months for a total of seven times during the study period of 18 months from February 1959 to July 1960. The same students in each school were tested on the same day throughout the survey using a single instrument for each function test measured by the same investigator. The Wright's peak flow meter is a handy small instrument widely used in epidemiological field study as an indicator of airway obstruction. The flow rate was taken as the average of the two highest among three readings for our own standardization. The instrument was dried frequently, after about every 59 maneuvers, using a nonheated hand-held hair dryer. Calibration was made each test day by both the investigator's own maximum forced flow rate and fixed steady air flow through the rotameter (a flow measuring instrument). Initial readings in January 1959 were discarded from analysis due to the students' lack of familiarity with procedures. Thereafter, students of age 10 to 11 easily got used to the physiological tests and thoroughly cooperated throughout the study. One complete test took about 2 min. The $FEV_{0.5}$ test was used as a supplementary test for comparison purposes to verify the peak flow rate's usefulness in air pollution study. The values of all the pulmonary function tests were converted to body temperature and pressure and saturated with water vapor (BTPS).

C. Routine Complaints of School Children

Rate of frequency of the routine perceptive complaints due to air pollution of the H- and K-School children under study in July 1959 is shown in Figure 3. In H-School, soiled clothes and books were the greatest complaints. Irritative sensations and cough and phlegm accounted for about one third. These complaints were suspected to be due to the effects of heavy suspended dust composed largely of relatively coarse particles. On the other hand, very low frequencies were shown for all the items at K-School. However, all children in both schools were recorded as healthy in annual clinical check-ups including X-rays by school doctors.

V. ANALYSIS

A. Anthropometric Measurements

The height and weight of the students of the two schools were measured six times in the course of the study from February 1959 to June 1960. The mean SD and number of subjects

Table 1

MONTHLY WIND DIRECTION, DUSTFALL, AND SULFATION RATE AT H-SCHOOL AND K-SCHOOL
(KAWASAKI CITY FEBRUARY 1959 to AUGUST 1960)

	1959											1960							
	Feb.	March	April	May	June	July	Aug.	Sept.	Oct.	Nov.	Dec.	Jan.	Feb.	March	April	May	June	July	Aug.
Dominant wind direction	NNE	NE	SSW	SSW	SEE	SE	SSE	NW	NW	NWW	NW	NWW	NE	NE	SSW	SSE	SSE	SSE	SW
Dustfall (ton/km²/month)																			
H-School[a]	15.4	24.0	70.0	50.2	44.8	43.3	50.0	42.5	38.2	30.4	24.0	15.7	30.0	36.8	30.6	44.0	45.8	51.5	61.3
K-School[b]	7.0	7.6	12.2	11.1	10.0	8.0	14.0	10.0	7.0	7.0	9.5	8.0	12.6	11.5	9.8	10.5	6.8	10.0	11.3
Sulfation rate (SO₃ mg/day/100 cm² PbO₂)																			
H-School[a]	0.85	0.90	1.70	1.65	0.90	1.00	1.55	0.80	0.70	0.46	1.95	0.48	0.55	0.85	0.82	1.70	1.92	1.80	1.48

[a]   Industrial polluted area.
[b]   Residential less polluted area.

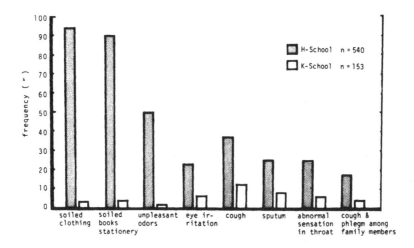

FIGURE 3. Frequency distribution of daily complaints of school children in polluted and less polluted areas.

measured each month for the cohorts are tabulated by age, sex, and school grade in Table 2. As stated above, the students' grades at the time of the first entry in February 1959 were the 4th and 5th because the school year in Japan begins in April and ends in March. Both height and weight gained gradually every month with the linearity of normal growth. It was noted that there was no statistical difference ($p < 0.05$ by Students' t-test) in the mean for each month between H- and K-School children by both sexes. This finding might allow a reasonable comparison of pulmonary function in each month without adjustment for body stature, if there was no serious attrition of the subjects, or if the correlation proved of sufficient magnitude to show association of cause and effect.

B. Vital Capacity

The monthly variations of the mean of unforced vital capacity of the two school children by a water-filled spirometer are shown in Table 3. During the 18 months the measurements were made seven times at H-School and three times at K-School, and a parallel gradual increase was observed as the ages of the children increased. In April 1959, February, and July 1960, no statistical differences were found between the two groups in the respective months. This could be taken to indicate that the vital capacity of the H-School students was not influenced acutely despite the sharp rise of dustfall and sulfur oxides.

C. Wright's Peak Flow Rate (PFR)

The follow-up data of Wright's peak flow rate of the exposed (H-School) and referent (K-School) students during the study period are shown in Table 3. In general, the values were increasing with their growth. It was noted, however, that there was a significant decline of the peak flow rate of H-School students during the polluted season in 1959 and a recovery in unpolluted periods and again a drop in polluted months in 1960, in contrast to no such variations in the K-School students throughout the study period. A statistically significant difference between the mean of the rate for both sexes was found in April and June in 1959 and for males in July in 1960 ($p < 0.05$, Students' t-test). Although data for the summer vacation were lacking, this finding suggested that the heavy air pollution affected the children's pulmonary ventilatory capacity temporarily and reversibly. Evaluation of PFR together with vital capacity data suggest that there might be sufficient exposure to dustfall and sulfur oxides to induce some degree of acute airway obstruction. The data for the 0.5-sec forced expiratory capacity in Table 3 indicated the same trends as do the Wright's peak flow rate.

Table 2
HEIGHT AND WEIGHT OF CHILDREN SELECTED AS COHORTS IN H-SCHOOL AND K-SCHOOL, KAWASAKI CITY DURING FEBRUARY 1959 TO JUNE 1960[a,b]

		1959				1960	
		Feb.	April	June	Nov.	Feb.	July
Height (cm)							
Male	H-School	131.6 ± 4.6(48)	132.8 ± 4.6(48)	133.8 ± 4.6(50)	135.1 ± 5.2(48)	136.7 ± 5.1(47)	138.4 ± 5.3(48)
	K-School	131.0 ± 4.5(48)	—	132.9 ± 4.7(48)	135.4 ± 5.0(49)	136.0 ± 5.0(49)	138.0 ± 4.8(50)
Female	H-School	132.2 ± 5.8(68)	133.5 ± 5.5(66)	134.8 ± 5.5(66)	137.1 ± 6.4(65)	138.8 ± 6.0(65)	140.9 ± 6.2(65)
	K-School	133.5 ± 6.0(45)	—	134.3 ± 6.9(43)	137.9 ± 6.6(43)	139.1 ± 6.5(43)	140.9 ± 5.7(44)
Weight (kg)							
Male	H-School	27.1 ± 3.3(48)	—	27.8 ± 3.3(50)	31.2 ± 3.6(48)	32.6 ± 3.6(47)	32.2 ± 3.9(48)
	K-School	27.5 ± 3.0(48)	—	28.4 ± 2.8(48)	30.0 ± 3.9(49)	32.0 ± 3.4(49)	31.3 ± 3.3(50)
Female	H-School	27.8 ± 4.1(68)	—	28.8 ± 4.2(66)	31.9 ± 4.9(65)	33.8 ± 5.2(66)	33.8 ± 5.5(65)
	K-School	28.0 ± 3.9(45)	—	29.6 ± 4.4(43)	30.5 ± 5.1(43)	31.2 ± 5.2(43)	33.9 ± 4.7(44)

Note: In each month, differences between H-School and K-School of both sexes are not significant ($p > 0.05$, student's *t*-test) in both height and weight.

[a] 5th graders 10—11 years old in June 1959.
[b] Mean ± SD (number of subjects).

Table 3

VITAL CAPACITY, WRIGHT PEAK FLOW RATE AND FEV$_{0.5}$ OF COHORT CHILDREN IN H-SCHOOL IN POLLUTED AREA AND K-SCHOOL IN LESS POLLUTED AREA, KAWASAKI CITY FEBRUARY 1959 TO JULY 1960

	1959 Feb.	1959 April	1959 June	1959 Nov.	1960 Feb.	1960 May	1960 July
Vital capacity (ℓ)							
H-School Male	1.93 ± 0.040(48)	1.99 ± 0.038(48)	1.97 ± 0.043(50)	2.00 ± 0.032(48)	2.05 ± 0.036(47)	2.15 ± 0.037(48)	2.11 ± 0.036(48)
Female	1.82 ± 0.040(68)	1.82 ± 0.042(66)	1.85 ± 0.038(66)	1.92 ± 0.037(67)	1.97 ± 0.047(65)	2.05 ± 0.040(65)	2.01 ± 0.039(65)
Mean	1.87 ± 0.033(116)	1.89 ± 0.032(114)	1.90 ± 0.029(116)	1.96 ± 0.029(113)	2.01 ± 0.032(112)	2.09 ± 0.030(113)	2.09 ± 0.029(113)
K-School Male	1.98 ± 0.038(50)				2.08 ± 0.037(49)	2.15 ± 0.033(50)	2.15 ± 0.033(50)
Female	1.83 ± 0.042(43)				1.94 ± 0.041(43)	1.94 ± 0.044(44)	1.94 ± 0.044(44)
Mean	1.90 ± 0.030(93)				2.01 ± 0.031(92)	2.06 ± 0.031(94)	2.06 ± 0.031(94)
Wright peak flow rate (ℓ/min)							
H-School Male	274 ± 7.9(48)	260 ± 7.4[a](48)	262 ± 6.6[d](50)	282 ± 6.1(48)	301 ± 6.9(47)	301 ± 7.2(48)	304 ± 7.1[g](48)
Female	273 ± 7.0(68)	267 ± 6.3[b](66)	270 ± 6.4[c](66)	287 ± 6.6(67)	306 ± 6.3(65)	309 ± 6.9(65)	312 ± 6.6(65)
Mean	274 ± 4.5(116)	264 ± 4.2(114)	266 ± 5.5[f](116)	286 ± 4.3(113)	305 ± 3.9(112)	305 ± 4.3(113)	308 ± 4.2(113)
K-School Male	275 ± 5.5(48)	280 ± 5.6[a](50)	290 ± 5.2[d](49)	300 ± 5.6(49)	311 ± 6.2(49)	320 ± 6.3(50)	324 ± 5.9[g](50)
Female	274 ± 5.0(45)	282 ± 5.2[b](43)	287 ± 4.5[c](43)	302 ± 5.2(43)	302 ± 5.0(43)	315 ± 4.8(43)	325 ± 4.4(44)
Mean	275 ± 3.6(93)	281 ± 3.7[c](93)	288 ± 3.4[f](92)	301 ± 3.5(92)	310 ± 3.7(92)	319 ± 3.6(93)	324 ± 3.8(94)
FEV$_{0.5}$ (ℓ)							
H-School Male	1.62 ± 0.037(47)	1.54 ± 0.030(47)	1.60 ± 0.025(49)	1.60 ± 0.027(48)	1.71 ± 0.024(47)	1.71 ± 0.032(48)	
Female	1.54 ± 0.036(66)	1.50 ± 0.032(65)	1.56 ± 0.024(65)	1.61 ± 0.027(66)	1.68 ± 0.020(65)	1.76 ± 0.030(65)	
Mean	1.59 ± 0.020(113)	1.52 ± 0.019(112)	1.57 ± 0.019(114)	1.61 ± 0.018(114)	1.69 ± 0.015(112)	1.74 ± 0.019(113)	

Note: Statistical difference between H- and K-Schools by student's *t*-test in each month; vital capacity — not significant ($p > 0.05$) for both sexes and their mean throughout April 1959 to July 1960. Mean ± standard error (number of subjects). Wright peak flow rate — significant ($p < 0.05$) for male,[a,d] female,[b,c] and their mean[c,f] in April and June 1959, and for male [g] in July 1960. Not significant in all other months for both sexes and mean.

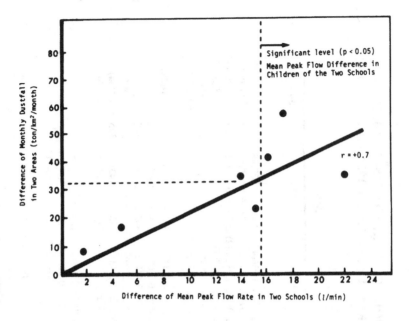

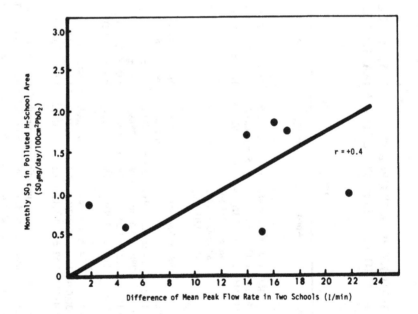

FIGURE 4.   Correlation between air pollution and Wright peak flow rate. (From Toyama, T., *Arch. Environ. Health,* 8, 165, 1964. With permission.)

The two measurements were highly correlated ($r = +0.7$) which proves the usefulness of either instruments for an air pollution survey.

D. Cause and Effect

The evidence that reversible fluctuation of peak flow rate occurred synchronized with seasonal variation of air pollution, and that the difference of the peak flow rate in the two groups became larger as the severity of air pollution increased indicates an association of cause and effect. The coefficient of correlation between the difference of the mean peak flow rate of the school children and the difference of dustfall or the level of sulfation rate in H-School for various months was r = +0.7 for dustfall and r = +0.4 for sulfation rate respectively as plotted in Figure 4. This indicated that the larger the difference of pollutant level, the larger the difference of the peak flow rate in the two groups. It can be roughly estimated that when the dustfall at both schools differed by more than 30 tons/km^2/month, the difference of mean peak flow rate between them became statistically significant ($p<0.05$) as shown in the same Figures. This might be accounted for by the Wright's peak flow rate being a rather less sensitive method to study the lower level of air pollution found at present.

VI. IMPLICATIONS AND CONCLUSION

As described above, this small scale study on health effects of air pollution in communities of limited area demonstrated a cause and effect association to which some of the epidemiological principles such as consistency, temporality, strength, and gradient of effect were to be met. However, there were some further points which might have been considered in the study's design and analysis. Although in a strict comparative study matched subjects should be paired, especially in the case of a cross-sectional study, repeated measurement in school grade-matched cohorts in this study might lead to a strong enough association. The monthly mean values of pulmonary functions should have been adjusted by body stature in comparing two groups. The conclusion using the crude mean in this study might be reasonably supported by the data of vital capacity which showed a trend of steady growth without difference in the two groups each month. The study period should have been longer than 2 years to provide more definite validity for seasonal variation. Because the result was urgently needed, the study was terminated after $1^1/_2$ years. Although no epidemics such as influenza were reported during the study period, other modifying factors which might cause airway obstruction such as temperature or humidity should have been evaluated. The subjects who did not react to the air pollution should also have been investigated. Regarding the statistical analysis, evaluation for variables of pollution and effects could have been made by a more sophisticated method like multivariate regression analysis. However, the evidence proved a cause and effect relationship with a reasonable degree of certainty even by using simple methods such as the *t*-test.

Figure 5 gives an overview of the pollution and its pulmonary effects in the course of this study. The results may indicate that the combined effects were induced by joint agents, sulfur oxides, and suspended particulate matter independently, additively, or synergistically.

A. Subsequent and Confirmatory Studies

To prove the study's replicability and generalizability, a similar health survey of children in Tokyo was undertaken in the following year, 1961, by a joint study team of the National Institute of Public Health and Keio University.[2,9] Children in six schools exposed to different levels of dustfall and sulfur oxides were followed up for 5 months. The mean Wright's peak flow rate of the children in polluted industrial areas fluctuated from month to month showing lower values than the children in clean residential areas. Here again, total vital capacity did not show any significant differences among the schools.

The decline of the Wright peak flow rate might be largely dependent on events in the large bronchial airways with inner caliber of more than 2 mm, which means that larger coarse particles in dustfall might react to large airways acutely and reversibly, while fine particles of dustfall reach the small airways and alveoli which develop chronic irreversible diseases.[10] The fact that unforced vital capacity did not change might indicate less effect on the small airways and alveoli.[11]

The results obtained in the Kawasaki study were not only a simple indication of temporary physical burden caused by pollutants. They signal also adverse physiological effects leading to the possibility of future chronic obstructive diseases such as chronic bronchitis in sensitive persons or susceptible populations.

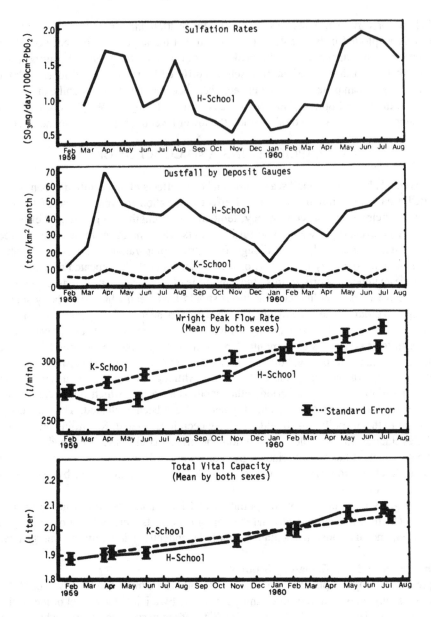

FIGURE 5.   Air pollution and pulmonary functions of school children in polluted H-School and less polluted areas, Kawasaki City, 1959-1960. (From Toyama, T., *Arch. Environ. Health*, 8, 165, 1964. With permission.)

B. Implications for Pollution Control

In the 5 years after the study, almost all large factories in the Kawasaki industrial area converted their fuel to oil and equipped themselves with dust precipitators and desulfurization systems. The dusty pollution then dropped sharply as shown in Figure 1. At present, 20 years after the study, the reducing type of air pollution has changed to a more diffuse oxidizing type with high levels of carbon monoxide, nitrogen oxides, and oxidants covering a wide area. Since about 1980, civil law suits have been brought by patients in epidemics of "compensatable environmental diseases" in the Kawasaki and Tokyo area. It is, however, difficult to determine whether their diseases are directly related to former pollution in Kawasaki, all this despite the results of this study which should have been useful for prevention of environmental diseases.

REFERENCES

1. Committee on Air Pollution, *Interim Report*, Her Majesty's Stationary Office, London, 1953.
2. **Toyama, T.,** Air pollution and its health effects in Japan, *Arch. Environ. Health*, 8, 153, 1964.
3. **Toyama, T. and Tomono, Y.,** Pulmonary ventilatory capacity of school children in a heavily air-polluted area, *Jpn. J. Publ. Health*, 8, 659, 1961.
4. **Huber, T. E., Joseph, S. W., Knoblock, E., Redfearm, P. L., and Karakawa, J. A.,** New environmental respiratory disease (Yokohama Asthma), *Arch. Ind. Hyg. Occup. Med.*, 10, 399, 1954.
5. D.S.I.R., *Measurement of Air Pollution*, Her Majesty's Stationary Office, London, 1957.
6. **Toyama, T.,** Air pollution, in *Japanese Medical Care Yearbook*, Jpn. Med. Assoc., Tokyo, 1968, 126.
7. **Katz, M.,** Some aspects of the physical and chemical nature of air pollution, in *Air Pollution*, World Health Organization, Geneva, 1961, 129.
8. **Wright, B. M. and MeKerrow, C. B.,** Maximum forced expiratory flow rate as a measure of ventilatory capacity, *Br. Med. J.*, 2, 1041, 1957.
9. **Suzuki, T., Ishikawa, K., Yokoyama, E., Kita, H., Maeda, H., Toyama, T., and Nakamura, K.,** Pulmonal-ventilatory capacity of school children in a city evaluated by Wright's peak flow rate, *Jpn. J. Ind. Med.*, 5, 525, 1963.
10. **Macklem, P. T. and Mead, J.,** Resistance of central and peripheral airways measured by a retrograde catheter, *J. Appl. Physiol.*, 25, 159, 1967.
11. **Bates, D. V.,** Air pollutants and the human lung, *Am. Rev. Resp. Dis.*, 105, 1, 1972.

Chapter 9

CRITICAL ASSESSMENT OF EPIDEMIOLOGIC STUDIES OF ENVIRONMENTAL FACTORS IN ASTHMA

Inge F. Goldstein and Diana Hartel

TABLE OF CONTENTS

I. INTRODUCTION

This chapter describes a series of interrelated epidemiologic studies of the role of air pollution in provoking symptoms in patients with asthma as well as new developments in research methods designed to overcome many of the drawbacks of traditional approaches to this problem. The advantages and pitfalls of past strategies employed in investigations of acute health effects of environmental agents as well as attempts to develop new strategies in asthma epidemiology are illustrated by a number of studies carried out with a nonwhite, low income, inner–city population with a high prevalence of asthma residing in New York City.

A. Magnitude of the Problem

The American Thoracic Society has defined asthma as "a disease characterized by an increased responsiveness of the trachea and bronchi to various stimuli and manifested by a widespread narrowing of the airways that changes in severity either spontaneously or as a result of therapy."[1] Asthma is a complex heterogeneous disease, the biological mechanisms of which are as yet unknown. The illness is chronic with episodic manifestations of symptoms, usually wheezing, difficulty in breathing, fatigue, cough, and tightness in the chest. Persons with asthma constitute approximately 2 to 3% of the general population in the U.S.[2,3] with 9 to 10 million affected people which, according to a 1975 study, incurred an estimated $629 million in direct medical costs.[4] While mortality from asthma is infrequent, its chronicity and severity make it a major public health problem.

The episodic appearances of asthma symptoms and attacks, resulting in significant interruptions in the lives of asthmatics and their families, are seldom witnessed in public. Consequently, there is little public awareness of the gravity of this illness. A glimpse of the personal suffering resulting from asthma is provided in the family records of Theodore Roosevelt. Although known for robust appearance and daring behavior as an adult, Roosevelt suffered from severe asthma as a child. Family letters and diaries detail the frequent attacks and the deep concern of his parents. His mother writes in October of 1869, "Teddie...seems hardly to have 3 or 4 days complete exemption and keeps us constantly uneasy and on the stretch."[5] At the time of that writing, attacks of asthma had been disturbing the Roosevelt household for 7 years and they continued until Roosevelt reached adulthood.

Research into the etiology of asthma dates from the time of Hippocrates; however, mechanisms of the disease process are still not well understood. An NIH panel has recently ascertained that "little is known about risk factors that are related to the prevalence, severity and persistence (chronicity) of asthma".[6] Efforts of asthma investigations, clinical as well as epidemiologic, have yielded some data on precipitants of acute episodes. Apart from occupational asthma, those environmental factors which have been proposed as precipitants can be roughly divided into several broad categories, the interaction between which may be additive or synergistic (i.e., more than additive):

1. *Allergens* such as those produced by cockroaches,[7,8] house dust mites,[9,10] mammals and birds,[11,12] molds and fungi,[13,14] pollen,[15,16] and food and drugs.[17,18]
2. *Irritants* such as SO_2,[19-21] ozone,[22,23] NO_2,[24,25] and mixtures of pollutants consisting of ozone, CO, and NO_2[26,27]
3. *Climatic factors* such as temperature (more specifically, sudden changes in temperature) and humidity,[20,29] and temperature inversions.[30]
4. *Respiratory infections.*[32-34]

In addition to external agents, exercise,[34,35] and emotional stress[36] have been shown to induce asthma symptoms as well as interact with any of the above precipitants of environmental origin. In this chapter, air pollutants cover both manmade and naturally occurring substances such as allergens found in the ambient air. The central question addressed in the study of the effects of air pollutants is whether acute manifestations, attacks as well as less severe narrowing of air passages and other symptoms, are related to irritant and/or allergen exposures. Various clinical and epidemiologic studies have been designed and executed to address the above question. They will be described in following sections to illustrate the major methodological approaches that have been taken in the investigation of acute health effects of environmental agents. The advantages and disadvantages of each study design will be discussed.

B. Clinical Studies

Clinical investigations of precipitants of asthmatic symptoms can be carried out by utilizing a controlled dose of the suspected agent either via inhalator, face mask, or in a chamber. One or more levels of exposure may be selected or the subject may be exposed to increasingly higher doses until a response is observed. Because of ethical considerations, such studies are usually performed with healthy adults or asthmatics who have mild cases of the disease, and the exposures used are generally of short duration. Although frequently used as a basis for air pollution standards, these studies do not represent exposures in the natural living environment for several reasons. First, the natural environment provides exposure to multiple contaminants which may have additive or synergistic effects. In addition, individuals may be exposed to short-term peak concentrations many orders of magnitude greater than the levels used in the clinic or laboratory. Effects in high-risk persons, such as those with more serious illness, the elderly, and children, remain unknown.

Despite the limitations of this approach, clinical studies can provide valuable clues to understanding and designing studies to investigate community disease patterns. Chamber studies have demonstrated increased airway reactivity among asthma patients to SO_2, NO_2, ozone, and pollutant mixtures.[19-27] Response to bronchial challenge with cockroach antigen,[37] and house dust (a complex mixture containing numerous allergens such as the house dust mite)[38] have been demonstrated as well. Several investigations[19-21] have corroborated a long-suspected greater sensitivity of asthmatics to air pollutants by demonstrating greater responsiveness in asthmatics when compared to nonasthmatics on exposure for short periods to SO_2 levels of 0.5 ppm. These SO_2 studies have become the basis for a newly proposed and more stringent air pollution standard for SO_2 proposed by the U.S. Environmental Protection Agency.[39] The USEPA has been mandated to protect the health of the entire U.S. population including the most sensitive members of the population.

C. Epidemiologic Investigations

The first observations of the effects of air pollution exposure on populations were of marked increases in mortality and morbidity, including many reports of increased adverse health effects in asthmatics, during well-documented air pollution disasters in the Meuse Valley, Belgium in 1930,[40] in Donora, Pa. in 1948,[41] and in London, England in 1952 and 1962.[42,43] In Donora, for example, the Public Health Service reported that 87% of the known asthmatics in the area had experienced respiratory symptoms during the 3-day severe smog episode.

Since the exact composition of pollutants in these disasters is not known, no definitive conclusions can be reached regarding agents responsible for the effects on asthma patients. Those pollutants whose concentrations were measured, however, reached levels that were more than tenfold greater than those found in ambient air today. These early reports mark the beginning of investigations into the effects of and governmental regulation of air pollution. In addition, findings of greater responsiveness of persons with asthma during severe smog episodes led to a belief that they constitute a pollution-sensitive population. This, however, has not been verified conclusively in epidemiologic studies.

The most common epidemiologic approach to assess both the acute and chronic health effects of air pollution on asthmatics is referred to by epidemiologists as the "ecological" study. Such a study is defined as an investigation in which the occurrence of an adverse health effect in a population is related to hypothetical causal agents to which a population as a whole is exposed without assessing individual exposures to those agents.[44,45] An example of such a study would be one in which population exposure is estimated from air pollution measurements at outdoor centrally located aerometric stations. Since such a study does not permit an evaluation of individual exposures which may differ from the average, it is incapable of demonstrating that the deleterious health effects are greatest among the more exposed individuals in the population. In air pollution and asthma research, two primary categories of studies have been carried out.

1. *Geographic* studies compare the occurrence of asthma cases or symptoms in an area of high pollution with that of an area of low pollution. In this approach, it is assumed that within each population individual exposures are relatively uniform and that the agents examined constitute the major difference between the two populations and can account for the difference in disease occurrence. These assumptions are rarely met in asthma research for the major reason that asthma is a heterogeneous condition with multiple precipitants which may differ for each asthmatic. In geographic studies, areas compared may have populations whose members differ greatly by type and degree of sensitivity to environmental agents. In pollution research in general, uniform exposure to air pollutants is highly unlikely in most populations as individuals differ by time spent in various pollution-related activities. Work in certain occupations, for example, can produce exposure to air pollutants many orders of magnitude greater than the average for the population as a whole. Geographic studies assume that such differentially exposed subgroups do not account for the observed differences in health effects between the compared locations. There are a number of geographic studies[47] which suggest an increased prevalence of asthma in areas of generally high pollution over areas of low pollution, although many counterexamples may also be found.

2. *Temporal* investigations relate changes over time in environmental factors to changes in the frequency of asthma attacks or symptoms. Increases in asthma visits to emergency rooms and increases in hospital admissions for asthma, indicating an elevation in asthma attack rates in populations, have been reported and related to changes in the levels of air pollutants in several urban populations, for example: Brisbane, Australia,[48,49] The Netherlands,[50] Yokkaichi, Japan,[51] New Orleans,[52-54] Los Angeles,[55] and New York City (see Section II. on asthma in New York City). Most communities around the world in which the temporal patterns of asthma have been studied show a definite rise in asthma attacks and symptoms in the fall season and many show another increase in the spring.

D. Methodologic Problems in the Epidemiologic Study of Acute Health Effects of Air Pollutants

There are a number of deficiencies in epidemiologic research into the health effects of air contaminants in general and in studies of asthma in particular. These deficiencies are categorized and described as follows:

1. *Poor quantification of exposure.* Air pollution measures often do not represent exposures of the population. Most epidemiologic investigations use data from one central aerometric station to represent exposure within a city or even larger areas. In New York City, for example, although it has an aerometric network of 40 stations, most investigators used data from only one central station to represent the day to day variation in levels of air pollution throughout the five boroughs of the city comprising an area of 320 square miles, some of which are located next to waterways or ocean and some not.[56,57] Changes in the levels of

air pollutants measured by one station were then related to fluctuations in mortality or morbidity of the population of 8 million residing in these five boroughs. It has been demonstrated, however, that levels of air pollutants measured in a central station in New York City do not represent the fluctuations or the absolute levels of the same contaminants in other areas of the city.[56-58] The correlation between daily levels of SO_2 or particulates measured at any two of the 40 stations of the aerometric network, for example, range between -0.9 and $+0.9$ and average approximately $+0.45$.[56] It has further been shown that the individual aerometric stations poorly represent levels of air pollution prevailing in areas in which they are situated.[57] In addition, exposure to indoor air pollutants has rarely been taken into consideration when estimating exposure to air pollutants in spite of the fact that most individuals spend 75 to 90% of their time indoors[59] and that levels and mixes of pollutants are known to differ in the indoor environment. This consideration is becoming more significant with recent measures to conserve energy by decreasing ventilation in homes and thus concentrating pollutants from indoor sources.

2. *Poor selection of exposure variables.* Sulfur dioxide and particulates are the most frequently used measures of pollution exposure, often because they are easy to monitor and are reported by most monitoring stations. The biological plausibility of the effects of these commonly measured pollutants are often derived from animal studies which, in addition to problems of extrapolating results from different species, generally use high exposure levels. Chemical or physical characterization, including size of particulates, is not usually specified nor are products of oxidation of SO_2 often included as measures of exposure. These compounds are suspected to have adverse health effects, but are rarely included in epidemiologic studies.

3. *Reliability of air pollution measurement.* Reliability (reproducibility) of measurements as reflected in side-by-side measurements of pollutants such as SO_2, for example, is frequently very low. The pararosaniline method commonly used in aerometric stations has been found to be unable to detect differences of 100% at a level of 100 μg/mg, a level commonly found in urban air, when results from two laboratories analyzing a single sample are compared.[60]

4. *Lack of data on peak exposures.* A few extreme peak concentrations may produce health effects which outweigh all other exposures, yet most pollution investigations employ averaged exposure data (24-hr averages or longer).

5. *Failure to account for trends.* Daily, weekly, or seasonal trends in both the independent (pollution) and dependent (health effect) variables may lead to the appearance of a spurious relationship that is accounted for by the effects of a third variable or set of variables (confounders), such as day of the week or climate.

In spite of some of the above-mentioned drawbacks, ecological studies are widely used because data are often available or relatively less expensive to acquire, and they serve as a satisfactory first approach in a search for environmental factors likely to cause health effects and to provide testable hypotheses for research.

II. EPIDEMIOLOGIC ASTHMA RESEARCH IN NEW YORK CITY

Epidemiologic investigations of asthma in New York City began with studies by Greenberg et al.[61] who found no increase in emergency room visits for asthma during a period of intense and prolonged air pollution in November 1953. In 1964 and 1965, these investigators, reported a seasonal trend in asthma visits to emergency rooms in the fall after the end of the pollen seasons, with "abrupt increases in the number of asthma visits on the same days at three of the hospitals studied".[62,63] Levels of air pollutants measured at a central aerometric station were not related to levels of emergency room visits for asthma at the above three hospitals, nor were outdoor pollen and mold counts related to these increases in asthma visits.

As previously pointed out, pollution measured at a single centrally located station was found to be a poor measure of city-wide exposure.[56–57] Demonstration of the poor reliability of what was then a standard approach to quantification of air pollution in health studies cast doubts on the validity of much previous work, and stimulated investigation into improving exposure assessment.

An alternative approach to the investigation of environmental influences on health indicators was consequently developed.[64] In this approach, the first step is to examine the chosen morbidity or mortality indicator so as to identify temporo-spatial patterns. These patterns can then serve as a screen to identify environmental conditions potentially related to the health events, which must conform to the temporo-spatial pattern of the health indicator. In an examination of acute asthma morbidity indicators, 9 years of emergency room data were gathered at three city hospitals, and a precise statistical method of identifying "epidemic" days was developed.[64,65] Days with unusually high numbers of asthma visits to hospital emergency rooms were defined as "epidemic" days if they exceeded a 15-day moving average of daily visits centered about the day in question by an amount felt to be statistically significant. This period of time was deemed small enough to adjust for the seasonal trend, and yet provide a sufficient number of days to estimate the random background of visits for that period.

It was found first, that during the autumn season in New York City there were a number of such days (asthma epidemic days), which, after taking into account the seasonal trend, exceeded chance expectation at a high level of statistical significance. Evidence supporting a role for environmental factors in asthma was found in that epidemic days tended to occur at different hospitals in different areas of the city simultaneously. This was highly suggestive of a predisposing environmental factor which, when present at all, was present over the entire city at one time.

This likelihood of a city–wide environmental factor or set of factors led to several attempts to link elevations in air pollutants to asthma epidemic days as well as days on which asthma visits to the emergency room were elevated. City–wide averages of SO_2 and particulate concentrations measured by the 40 station aerometric network were used for an estimate of air pollution levels, and several different statistical procedures were used to test for a relation between asthma visits and air pollution.[66] In Table 1, results of analyses comparing SO_2 and particulate levels on asthma epidemic days with levels on nonasthma epidemic days are presented. No tendency for the air pollution level to be high either on the day of the "epidemic" or on the previous day was found. In a second test, all days on which emergency room visits were significantly higher than normal (which includes epidemic days and other days with a less significant excess of asthma as well) were compared with days on which pollution levels were higher than normal. Again no relation was found. The statistical power of the test (the probability that a test will detect an effect) was examined and found to be sufficient to detect a reasonable effect, if one existed, of air pollution on asthma.

Because of the ecological nature of the study, these negative findings could not rule out the possibility that some patients with asthma are sensitive to the pollutants measured (sulfur dioxide and particulates) but were not necessarily exposed to the levels which prevailed in the ambient air. Negative findings in ecological research such as these cannot be construed to be evidence of the safety of contaminants at the levels examined; rather they simply indicate the lack of a detected effect given the methods employed.

In another approach, again using 9 years of daily visits for asthma to the three municipal hospitals in the innercity study areas, no trend was found over time in the weekly levels of emergency room visits: the seasonal patterns repeat themselves in a consistent way every year, with no significant changes occurring from one year to the next. Two years in particular

Table 1

TEST FOR DIFFERENCES IN MEAN POLLUTION LEVELS FOR (1) ASTHMA EVENTS VS. ALL OTHER DAYS AND (2) FOR A LAG EFFECT. EACH VALUE REPRESENTS THE MEAN ± SD (NUMBER OF OBSERVATIONS)

		(1)					(2)				
	Seasonal average	A days	Ā days	$t_{obs.}$	df	$t_{0.05}$	B days	B̄ days	$t_{obs.}$	df	$t_{0.05}$
SO₂: ppm × 10⁻³											
Fall '69	68.49 ± 19.69(60)	69.12 ± 26.64(5)	68.43 ± 19.26(55)	0.074	58	1.672	69.50 ± 24.29(8)	68.25 ± 18.58(47)	0.168	53	1.674
Fall '70	68.61 ± 18.56(60)	70.73 ± 20.04(7)	68.33 ± 18.54(53)	0.319	58	1.672	63.37 ± 10.22(5)	68.85 ± 19.19(48)	−0.626	51	1.676
Fall '71	34.19 ± 12.82(60)	30.03 ± 2.43(2)	34.33 ± 13.01(58)	−0.463	58	1.672	38.68 ± 24.72(5)	33.92 ± 11.69(53)	0.779	56	1.673
Smk: COHS/10⁴ ft²[a]											
Fall '69	15.13 ± 4.15(91)	12.78 ± 3.20(12)	15.48 ± 4.18(79)[b]	−2.140	89	1.662	13.86 ± 2.91(10)	15.72 ± 4.30(69)	−1.321	77	1.666
Fall '70	10.92 ± 3.90(91)	10.80 ± 3.38(12)	10.93 ± 3.99(79)	−0.107	89	1.662	8.71 ± 3.07(7)	11.15 ± 4.02(72)	−1.559	77	1.666
Fall '71	9.27 ± 2.64(91)	7.54 ± 1.75(9)	9.45 ± 2.66(82)[b]	−2.099	89	1.662	9.35 ± 3.44(7)	9.46 ± 2.60(75)	−0.104	80	1.665

Note: (1) All days are partitioned into A (asthma "event" days) vs. Ā (all other days). (2) All Ā days are partitioned into B (days immediately preceding A days) and B̄ (all other Ā days).

Test for A vs. Ā: H_0: $\mu_A = \mu_{\bar{A}}$, H_1: $\mu_A > \mu_{\bar{A}}$; H_0 is rejected (at $\alpha = 0.05$) if $t_{obs.} > t_{0.05}$, where $t_{obs.} = \bar{X}_A - \bar{X}_{\bar{A}}/S_{pooled} \sqrt{1/N_A + 1/N_{\bar{A}}}$. Similarly, for B vs. B̄.

[a] COHS = Coefficients of haze.
[b] The difference between asthma event days and nonasthma event days is significantly greater than $p < 0.05$ with less pollution on asthma event days.

were of interest — 1970 to 1971 when air pollution dropped drastically with declines of 80% in SO_2 and 60% in particulates as a response to the imposition of environmental standards. To examine a concomitant hypothesized change in asthma admissions, the ratio of asthma visits to emergency room visits for other respiratory conditions during this period was employed.[67] This ratio method was used in order to take into account a possible simultaneous decrease in asthma attack rate together with an increase in the population base using the hospital emergency rooms. The ratio of weekly visits for asthma to weekly visits for all other respiratory conditions did not change significantly during that period suggesting no effect on asthma.

The possibility that exposure to short-term peaks of an air contaminant was a more significant predisposing factor than averaged exposure estimates was addressed by examining excess asthma visits in relation to hourly levels of sulfur dioxide.[67] Data came from the 40-station New York City aerometric network which was in operation during the 1960s, through the early 1980s. During the 1960s, pollution frequently exceeded current standards and peaks included levels that have been shown to induce bronchoconstriction in laboratory studies of asthmatics.[19-21]No relationship between days when high hourly levels of sulfur dioxide were measured and days with high numbers of emergency room visits for asthma was found when compared to days without high hourly levels of SO_2.

When asthma epidemic days were compared with atmospheric conditions, a highly statistically significant association was found between asthma epidemic days and high-pressure conditions with low nocturnal wind speeds in New York City and also in New Orleans.[30] This association, however, does not necessarily imply a meteorological etiology, since high pressure systems tend to be associated with stagnant and rain-free conditions which favor the accumulation of airborne pollutants in the atmosphere. Nor does the association necessarily imply an etiologic agent in the ambient outdoor air, as air exchange rates between the indoor and outdoor environments are known to be reduced under calm atmospheric conditions, thus allowing the build-up of indoor contaminants. Although the specific agents responsible remain to be identified, the demonstration of association of asthma epidemic days with an atmospheric variable supports the hypothesis that environmental factors may be mediators of asthma attacks.

A. New York City and New Orleans

Comparison of asthma patterns in New York City with patterns in New Orleans revealed contrasting features which were unique to each area and thus likely to be due to local environmental conditions. Twenty-five years of emergency room visits to Charity Hospital in New Orleans were compared with data for a 9-year period (1969 to 1977) for New York City hospitals in Harlem and Brooklyn.[68] Unlike New York City, the "epidemic" days in New Orleans were not consistently concentrated in the fall season, but rather were distributed over the whole year, with a slight increase in the spring and the fall in some years but not in all years. Another difference appeared in seasonal trends of asthma visits. Every year the New York City hospitals had a steady rise in numbers of daily visits for asthma beginning in late September, reaching a peak in October and November, and declining in December, an effect over and above the clustering of epidemic days in the fall. New Orleans asthma often showed a seasonal rise in the fall but not the consistent increase found in every year in New York City. Representative yearly plots of daily visits for asthma in the two cities are presented in Figure 1. Note that the seasonal increase in New York City results from a gradual increase in the day-to-day total visits and not primarily from the greater numbers of epidemic days in that season. Such a pattern was found in all three hospitals for all 9 years of observation in New York City.

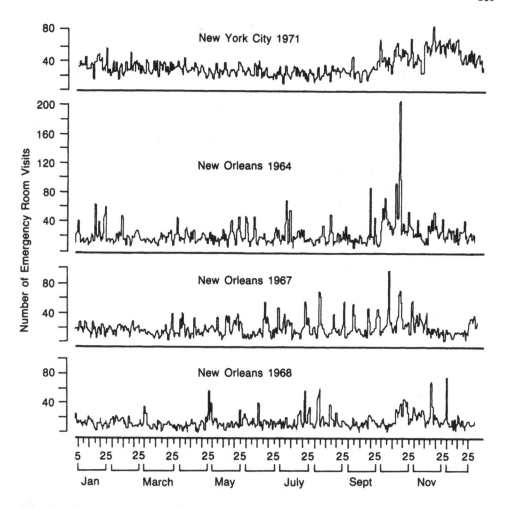

FIGURE 1. Daily emergency room visits for asthma for New York City (Cumberland, Kings County, and Harlem's hospitals combined) and New Orleans.

Still another difference concerned the day-of-the-week pattern in asthma[69] which differed drastically in the two cities. In New York City there was a marked excess of asthma attacks and of asthma epidemic days on Sundays and Mondays for the 9 years of observation in all 3 hospitals studied. In contrast, New Orleans had neither an excess of epidemic days on Sundays nor an excess of daily visits on Sundays; visits for asthma were consistently low on Sundays during all 25 years for New Orleans.

In order to rule out effects of daily patterns of emergency room use which might account for the occurrence of excess asthma visits on certain days of the week, total visits to emergency rooms for respiratory complaints and visits to emergency rooms for all causes excluding trauma were examined. The excess visits for asthma on Sundays contrasted sharply with visits for other causes in New York City which were lowest on Sundays and highest on Mondays. This excess of asthma attacks on Sundays in New York City suggests that in New York City some factor related to human activities, particularly in the fall season, triggers attacks. In New Orleans, which does not show this pattern, a different set of environmental conditions may be affecting most of that area's asthmatic population, most likely factors related to outdoor allergens. The more regular pattern in New York City on the other hand, which appears to have its origin in patterns of human activities, such as staying indoors at home on weekends, is likely to be predominantly related to an indoor exposure.

III. A SHIFT IN STRATEGY: INDIVIDUAL-CENTERED EPIDEMIOLOGY

To avoid the drawbacks of the ecological approach, and to obtain a better estimate of individual exposures, the focus of asthma and air pollution epidemiology has been shifted in our current investigations to the personal environment of individuals in inner-city areas of New York City. This approach examines changes in environmental factors on the level of the individual in relation to changes in his or her health status over time. Impetus to examine the personal environment of individuals was provided by the knowledge that most persons spend a great percentage of their time indoors and by evidence from the studies described above which led us to suspect the influence of an indoor factor in New York City asthma.

Investigations of indoor pollution have only recently received public attention, although many studies of the indoor environment have been made in disparate fields of study (infectious diseases, bioallergens, appliance evaluation, etc). A number of reviews [70-75] now exist which summarize these investigations. It has been observed that, due to the existence of sources of pollutants operating within the home, and also due to insulation measures leading to reduced air exchange as an energy conservation measure, pollutants may become entrapped in buildings.[76,77] Based on indoor contaminant research, the following generalizations may be made:

1.   Indoor concentrations of certain pollutants frequently exceed their ambient levels outdoors.
2.   Indoor exposures, both average levels and short-term peaks, often greatly exceed current outdoor air quality standards for certain contaminants.
3.   Some pollutant levels such as SO_2 are generally found to be lower indoors.

Of particular interest to epidemiology is the fact that apart from health care facilities, residences are the primary environment of high-risk persons — the elderly, the ill, and the very young — who often spend 100% of their day indoors.

A. The Indoor Environment in New York City

Specific findings in New York City regarding the indoor environment were first made in a sample of 120 inner-city residents in the South Bronx,[78] who were identified on the basis of utility company records as residing in an area in which gas stoves were heavily used. In many homes, levels of NO_2 and CO were found which exceeded outdoor concentrations and which were higher than EPA standards. It was found that 53% of respondents insulated their homes with a variety of materials and many of these brick and stone low-income homes were found to be very tightly sealed. In addition to insulation practices of residents, landlords often retrofitted tight windows to conserve heating fuel. A set of social and other conditions associated with residence in low income neighborhoods in New York City, furthermore, resulted in residents spending a great deal of time indoors: harsh winters, economic factors such as lack of work or inability to purchase entertainment and vacations outside the home, and the higher crime rate in poor areas which limited access to parks and other free local recreational activities.

A second investigation of the residential environment of asthmatic children is currently being conducted.[79] Preliminary results have amplified earlier findings of a high ratio of indoor to outdoor pollution and documented the varied exposure patterns which occur in each home. In addition, greatly reduced ventilation which allows build-up of irritants as well as allergens whose source is within the home has been documented. High temperature

Table 2
AVERAGE NUMBER OF HOURS[a] PER DAY SPENT BY STUDY SUBJECTS INDOORS AND OUTDOORS

Location	Fall 1982 Child (n = 17)	Fall 1982 Adult (n = 13)	Winter 1983 Child (n = 3)	Winter 1983 Adult (n = 9)	Summer 1983 Child (n = 9)	Summer 1983 Adult (n = 8)	Fall 1983 Child (n = 9)	Fall 1983 Adult (n = 3)
Home								
H[b]	20.1	20.6	21.9	17.4	18.9	15.5	18.9	18.6
NH[c]	17.3	19.4	19.0	15.0	18.1	15.3	15.9	16.9
Indoors (not home)								
H	1.8	1.8	0.7	4.3	1.5	3.1	3.2	2.8
NH	4.9	2.3	3.6	5.8	1.8	5.9	6.2	3.8
Outdoors								
H	2.1	1.6	1.4	2.3	3.6	5.4	1.9	2.6
NH	1.8	2.3	1.4	3.2	4.1	2.8	1.9	3.3

[a] Averages are based on daily activity questionnaires.
[b] H: Holidays include Saturday, Sunday, and public school holidays.
[c] NH: Nonholidays.

(75 to 80°F) and humidity, conditions which allow proliferation of bioallergens such as mold and house mites, were often encountered even during cold weather as residents were found to prefer a warm, humid indoor climate. That residents spent a high percentage of their time indoors, especially asthmatic children who were restricted either by their parents or by episodes of illness, was borne out in tallied results of daily activity diaries. Table 2 clearly demonstrates the high percentage of time (shown as average numbers of hours) spent by adults and children indoors, especially indoors at home.

B. Research Methods for Individual-Centered Studies

On the basis of these indoor pollution studies in New York City, it has become increasingly obvious that even if New York City's air monitoring network could adequately measure local outdoor air pollutants, it would be unable to assess exposures of persons with asthma living in low-income areas of New York City as these exposures are highly dependent on the conditions of the personal environment and on human activities. The great variability found in individual pollution exposure patterns, including short-term peak exposures, is difficult to predict without direct measurement in the personal environment. In asthma research, additional sources of individual variability derive from the large number of potential triggers of asthma symptoms and attacks with each asthmatic conceivably responding to a different set of environmental agents.

Given the highly individual nature of exposures in New York City and of responses in asthma, an appropriate study design would be to monitor individuals who would serve as their own controls over time. Repeat measures (time series) of exposure and health effect may be used in which comparisons are made on individuals against their own set of baseline responses when hypothesized triggers of asthma are absent. The same measures and study protocol are used for all persons enrolled in the investigation, and, as the number of individuals in the study increases, subgroupings may be made of persons with similar dose-response patterns.

There is currently great interest in the development of monitors for better estimation of short-term exposures in order to carry out studies of the personal environment. Samplers which can be worn or carried, for example, have been designed to detect contaminants such as CO, NO_2, SO_2, formaldehyde, various organics, and particulates, and for many, instrument sensitivity and specificity is sufficient for use in individual time series studies. Similar advances are being made in the measurement of possible indicators of subclinical disease. Lightweight portable instruments which measure lung function, for example, can provide a more objective estimate of daily change in pulmonary status than less reliable questionnaire data.[80,81]

In addition to the improvements in the measurement of agent and outcome on the level of the individual, statistical methods appropriate to analyze results of studies in which exposures of individuals over time is related to health indexes both on the individual level and on the level of groups have been developed and tested. In the past, the most common approach was to use regression techniques relating symptoms for a panel of subjects as a whole against a set of air pollution measures which are examined singly or in combination. This approach to using panels of subjects has serious drawbacks which have been pointed out by Whittemore et al.[82,83] These investigators have developed statistical methods for panel studies which use individual logistic regressions for each study member and which can then be used for the panel as a whole or to form subgroups of panelists. Data is collected in a method consistent with individual-centered study design: each person is an experiment over time and serves as his or her own control.

Whittemore and Korn tested their methods using EPA data for which an association was found between asthma attacks and total suspended particulates.[84] Although their statistical methods were mathematically innovative, the study was ecological and the investigators concluded that the following defects remained.

1.   Exposure data was based on a central city–wide monitoring system and probably did not represent individual exposures.
2.   Subjective reports of asthma attacks and symptoms by study participants were likely to be unreliable.

The methods, however, are extremely powerful when refinements in exposure and outcome are utilized and when the number of days for which individuals report symptoms becomes large enough.

IV. SUMMARY AND CONCLUSIONS

Epidemiologic investigations of the relationship between air pollution exposures and asthma attacks, beginning with pollution disasters in the first half of this century, have been presented. Methodological issues in asthma epidemiology have been discussed in particular in relation to valid estimation of exposure, measures of health effects, study design, and statistical methods. Of special interest is the development of an individual-centered approach to the study of asthma in New York City. The methodological advances described above in statistical analysis and pollution monitoring equipment have made possible and desirable this approach to acute health effects research, which combines features from both clinical and epidemiologic study designs. It is able to incorporate the desirable features of clinical studies, in which physiological measures are emphasized and features of epidemiologic studies, in which members of a population exposed to naturally occurring or selfselected agents are studied over an extended time period. While the exposure is not controlled as is usually the case in clinical studies, the use of personal monitors provides more precise estimation of individual exposures to relate to more objective and reliable measures of response such as found in the clinic. As in the clinic, each individual may serve as his or her own control, but a much longer observation period is used during which individual baseline data may be established and multiple exposures may be examined.

The importance of the personal environment in the assessment of air pollution exposure coupled with the heterogeneous nature of asthma makes the area of acute health effects research particularly well-suited to individual-centered epidemiology using technical developments in design, statistics, and monitoring equipment. The short time period, in most cases of asthma, between exposure and outcome (minutes to hours after exposure) makes this study design in asthma epidemiology particularly feasible, although the general principles have applicability to other areas of epidemiologic research. Application to chronic diseases with long latencies such as cancer can also be anticipated when advances in measurement of biological markers of exposure to carcinogens and early physiological responses are accomplished. For example, in prospective studies in occupational settings where workers undergo frequent periodic medical examinations and records of variables such as blood-cell counts starting from preemployment through employment can be related to their exposure history. The use of such biological markers is expected to improve understanding of the relationship between exposure to suspected human carcinogens and the ultimate development of cancers while taking individual susceptibility and natural history of disease into account.

REFERENCES

1. Respiratory Diseases: Task Force Report, DHEW Publ. (NIH 73-432, National Institutes of Health, Bethesda, Md., 64, 1972.
2. Report of the Task Force on Asthma and Other Allergic Diseases, National Institute of Allergy and Infectious Diseases, USDHHS, National Institutes of Health, Bethesda, Md., 1979.
3. **Gordis, L.**, *Epidemiology of Chronic Lung Diseases in Children*, Johns Hopkins Press, Baltimore, 1973.
4. **Young, P.**, Asthma and Allergies: An Optimistic Future, U.S. Department of Health and Human Services, NIH #80-382, Bethesda, Md., March 1980.
5. **McCullough, D. G.**, Mornings on horseback, *A Disease of Direct Suffering*, G. K. Hall, New York, 1981, chap. 4.
6. Epidemiology of Respiratory Diseases: Task Force Report July 1979, U.S. Department of Health and Human Services, Public Health Service, National Institutes of Health, NIH #81-2019, Bethesda, Md., October 1980.
7. **Kang, B., Vellody, D., Homburger, H., and Yunginger, J. W.**, Cockroach: cause of allergic asthma, its specificity and immunologic profile, *J. Allergy Clin. Immunol.*, 63, 80, 1979.
8. **Bernton, H. S., McMahon, T. F., and Brown, H.**, Cockroach asthma, *Br. J. Dis. Chest*, 66, 61, 1972.
9. **Murray, A. B., Ferguson, A. C., and Morrison, B. J.**, Diagnosis of house duct mite allergy in asthmatic children: what constitutes a positive history?, *J. Allergy Clin. Immunol.*, 71, 21, 1983.
10. **Voorhorst, R.**, Specific causes of bronchial asthma: mites and house dust, reprinted from the *Excerpta Med. Int. Congr. Ser.*, No. 232, October 1970.
11. **Fagerberg, E. and Wade, L.**, Diagnosis of hypersensitivity to dog epithelium in patients with asthma bronchiale, *Int. Arch. Allergy Appl. Immunol.*, 39, 301, 1970.
12. **Middleton, E., Reed, C. E., and Ellis, E. F.**, *Allergy, Principles and Practice*, C. V. Mosby, St. Louis, 1978.
13. **Frankland, A. W.**, Mold, fungi and bronchial asthma, in *Bronchial Asthma Mechanisms and Therapeutics*, Weiss, E. B. and Segal, M. S., Eds., Little, Brown, Boston, 1976.
14. **Salvaggio, J. and Aukrust, L.**, Mold-induced asthma, *J. Allergy Clin. Immunol.*, 68, 327, 1981.
15. **Bruce, L. A., Rosenthal, R. R., Lichtenstein, L. M., and Norman, P. S.**, Diagnostic tests in ragweed-allergic asthma, *J. Allergy Clin. Immunol.*, 53, 230, 1974.
16. **Novey, H. S., Roth, M., and Wells, I. D.**, Mesquite pollen, an aeroallergen in asthma and allergic rhinitis, *J. Allergy Clin. Immunol.*, 59, 359, 1977.
17. **Aas, K.**, Studies of hypersensitivity to fish, *Int. Arch. Allergy*, 29, 346, 1966.
18. **McDonald, J. R., Mathison, D. A., and Stevenson, D. D.**, Aspirin intolerance in asthma: detection by oral challenge, *J. Allergy Clin. Immunol.*, 50, 198, 1972.
19. **Linn, W. S., Bailey, R. M., Medway, D. A., Venet, J. G., Wightman, L. H., and Hackney, J. D.**, Respiratory responses of young adult asthmatics to SO_2 exposure under simulated ambient exposure conditions, *Environ. Res.*, 1982.
20. **Koenig, J. Q., Pierson, W., Horike, M., and Frank, R.**, Effects of SO_2 plus NaCl aerosol combined with moderate exercise on pulmonary function in asthmatic adolescents, *Environ. Res.*, 25, 340, 1981.

21. **Sheppard, O., Wong, W. S., Uehara, C. F., Nadel, J. A., and Boushey, H. A.,** Lower threshold and greater bronchomotor responsiveness of asthmatic subjects to SO_2, *Am. Rev. Resp. Dis.*, 122, 873, 1980.

22. **Hackney, J. D., Linn, W. S., Mohler, J. G., Pederson, E. E., Breisher, P., and Russo, A.,** Experimental studies on human health effects of air pollutants. II. Four-hour exposure to ozone alone and in combination with other pollutant gases, *Arch. Environ. Health*, 30, 379, 1975.

23. **Hackney, J. D., Linn, W. S., Mohler, J. G., and Collier, C. R.,** Adaptation to short-term respiratory effects of ozone in men exposed repeatedly, *J. Appl. Physiol.*, 43, 82, 1977.

24. **Orehek, J., Massari, J. P., Gayrard, P., Grimaud, C., and Charpin, J.,** Effect of short-term low-level nitrogen dioxide exposure on bronchial sensitivity of asthmatic patients, *J. Clin. Invest.*, 57, 301, 1978.

25. **Kerr, H. D., Kulle, T. J., McIlhany, M. L., and Swidersky, P.,** Effects of nitrogen dioxide on pulmonary function in human subjects: an environmental chamber study, *Environ. Res.*, 19, 392, 1979.

26. **von Nieding, G., Wagner, H. M., Krekeler, H., Lollgen, H., Fries, W., and Beuthan, A.,** Controlled studies of human exposure to single and combined action of NO_2, O_3 and SO_2, *Int. Arch Occup. Environ. Health*, 43, 195, 1979.

27. **Bell, K. A., Linn, W. S., Hazucha, M., Hackney, J. D., and Bates, D. V.,** Respiratory effects of exposure to ozone plus sulfur dioxide in southern Californians and eastern Canadians, *Am. Ind. Hyg. Assoc. J.*, 38, 696, 1977.

28. **Miller, J. S.,** Cold air and ventilatory function, *Br. J. Dis. Chest*, 59, 23, 1965.

29. **Goldstein, I. F.,** Weather patterns and asthma epidemics in New York City and New Orleans, *Int. J. Biometeor.*, 24, 329, 1980.

30. **Goldstein, I. F. and Raynor, G.,** Temporal patterns of asthma, *Int. Aerobiol. Newsl.*, 17, 23, 1983.

31. **Minor, T. E., Dick, E. C., Baker, J. W., Ouellette, J. J., Cohen, M., and Reed, C. E.,** Rhinovirus and influence type A as precipitants of asthma, *Am. Rev. Resp. Dis.*, 113, 149, 1976.

32. **Ellis, E. F.,** Relationship between the allergic state and susceptibility to infectious airway disease, *Pediatr. Res.*, 11, 227, 1977.

33. **Clarke, C. W.,** Relationship of bacterial and viral infections to exacerbations of asthma, *Thorax*, 34, 344, 1979.

34. **Balfour-Lynn, L., Tooley, M., and Godfrey, S.,** Relationship of exercise induced asthma to clinical asthma in childhood, *Arch. Dis. Child.*, 56, 450, 1981.

35. **Sheppard, D. A., Saisho, A., Nadel, J. A., and Boushey, H. A.,** Exercise increases SO_2 induced bronchoconstriction in asthmatic subjects, *Am. Rev. Resp. Dis.*, 123, 486, 1981.

36. **Luparello, T., Lyons, H. A., Bleecker, E. R., and McFadden, E. R.,** Influences of suggestion on airway reactivity in asthmatic subjects, *Psychosom. Med.*, 30, 819, 1968.

37. **Kang, F.,** Study on cockroach antigens as probably causative agent in bronchial asthma, *J. Allergy Clin. Immunol.*, 58, 357, 1976.

38. **Booij-Noord, H., de Vries, K., Sluiter, H. J., and Orie, N. G. M.,** Late bronchial obstructive reaction to experimental inhalation of house dust extract, *Clin. Allergy*, 2, 43, 1972.

39. Air Quality Criteria for Particulate Matter and Sulfur Oxides, Vol. 3, EPA-600/8-82-029c, U.S. Environmental Protection Agency, Washington, D.C., December 1982.

40. **Firket, J.,** Sur les causes des accidents survenus dans la valee de la meuse, lors des Brouillards de Decembre 1930, *Bull. Acad. R. Med. Belg.*, 11, 683, 1931.

41. Air pollution, in Epidemiology of the Unusual Smog Episode of October 1948, Donora, P. A., Public Health Bulletin 306, Washington, D.C., 1949.

42. Mortality and Morbidity during the London Fog of December 1952, Reports on Public Health and Related Subjects, No. 95, Ministry of Health, London, 1954.

43. **Fry, J., Dillane, J. B., and Fry, L.,** Smog — 1962 vs. 1952, *Lancet*, 2, 1326, 1962.

44. **Susser, M.,** *Causal Thinking in the Health Sciences: Concepts and Strategies of Epidemiology*, Oxford University Press, New York, 1973.

45. **Kleinbaum, D. G., Kupper, L. L., and Morgenstern, H.,** *Epidemiologic Research*, Wadsworth, Belmont, Calif., 1982.

46. **Lebowitz, M. D. and Burrows, B.,** Tucson epidemiologic study of obstructive lung diseases. II. Effects of in-migration factors on the prevalence of obstructive lung diseases, *Am. J. Epidemiol.*, 102, 153, 1975.

47. **Ferris, B. G. and Whittenberger, J. L.,** Environmental hazards: effects of community air pollution on prevalence of respiratory disease, *N. Engl. J. Med.*, 275, 1413, 1966.

48. **Derrik, E. H., Thatcher, R. H., and Trappert, L. G.,** The seasonal distribution of hospital admissions for asthma in Brisbane, *Aust. Ann. Med.*, 9, 180, 1960.

49. **Derrik, E. H.,** The seasonal variation of asthma in Brisbane: its relation to temperature and humidity, *Int. J. Biometeor.*, 9, 239, 1965.

50. **Tromp, S. W.,** Biometeorological analysis of frequency and degree of asthma attacks in western part of Netherlands, in *Proc. 2nd Bioclimatic Congress*, Macmillan, New York, 1962, 477.

51. **Yoshida, E., Hidehiko, O., and Immai, M.,** Air pollution and asthma in Yokkaichi, *Arch. Environ. Health*, 13, 763, 1966.

52. **Salvaggio, J., Hasselblad, V., Seabury, J., and Heiderscheit, L. T.,** New Orleans asthma. II. Relationship of climatologic and seasonal factors to outbreaks, *J. Allergy,* 45, 257, 1970.

53. **Carroll, R. F.,** Epidemiology of New Orleans asthma, *J. Am. Public Health,* 58, 1677, 1968.

54. **Lewis, R., Gildeson, M. D., and McCalden, R. O.,** Air pollution and New Orleans asthma, *Public Health Rep.,* 77, 947, 1962.

55. **Schoettlin, C. E. and Landau, E.,** Air pollution and asthmatic attacks in the Los Angeles area, 76, 545-548, 1961.

56. **Goldstein, I. F. and Landovitz, L.,** Analysis of air pollution patterns in New York City. I. Can one station represent the large metropolitan area?, *Atmospheric Env.,* 11, 47, 1977.

57. **Goldstein, I. F. and Landovitz, L.,** Analysis of air pollution patterns in New York City. II. Can one station represent the area surrounding it?, *Atmospheric Env.,* 11, 53, 1977.

58. **Goldstein, I. F., Landovitz, L., and Glock, G.,** Air pollution patterns in New York City, *J. Air Pollut. Control Assoc.,* 24, 148, 1974.

59. **Szalai, A.,** *The Use of Time: Daily Activities of Urban and Suburban Populations in 12 Countries,* Mouton, The Hague, 1972.

60. **Goldstein I. F. and Goldstein, M.,** Evaluation of research strategies for investigation of health effects of air pollution, *Bull. N.Y. Acad. Med.,* 54, 1119, 1978.

61. **Greenberg, L., Jacobs, M. B., Drolette, B., Field, F., and Braverman, M.,** Air pollution and morbidity in New York City, *JAMA,* 182, 161, 1962.

62. **Greenberg, L., Erhardt, C. L., Field, F., and Reed, J.,** Air pollution incidents and morbidity studies, *Arch Environ. Health,* 10, 351, 1965.

63. **Greenberg, L.,** Asthma and temperature change, *Arch. Environ. Med.,* May 1964.

64. **Goldstein, I. F. and Rausch, L.,** Time series analysis of morbidity data for assessment of acute environmental health effects, *Environ. Res.,* 17, 266, 1978.

65. **Goldstein, I. F. and Cuzick, J.,** Application of time-space clustering methodology to the assessment of acute environmental health effects, *Rev. Env. Health,* 111, 259, 1981.

66. **Goldstein, I. F. and Dulberg, E.,** Air pollution and asthma: search for a relationship, *J. Air Pollut. Control Assoc.,* 32, 370, 1981.

67. **Goldstein, I. F. and Weinstein, A.,** Air pollution and asthma: effect of exposure to short-term sulfur dioxide peaks, *Environ. Res.,* in press.

68. **Goldstein, I. F. and Currie, B.,** Seasonal patterns in asthma, clues to etiology, *Environ. Res.,* 33, 201, 1984.

69. **Goldstein, I. F. and Cuzick, J.,** Daily patterns of asthma in New York City and New Orleans: an epidemiologic investigation, *Environ. Res.,* 30, 211, 1983.

70. **Yocom, J. E.,** Indoor-outdoor air quality relationships, a critical review, *J. Air Pollut. Control Assoc.,* 32, 500, 1982.

71. **Committee on Indoor Pollutants,** *Indoor Pollutants,* National Academy of Sciences, National Academy Press, Washington, D. C., 1981.

72. **Sterling, T. and Kobayashi, D.,** Exposure to pollutants in enclosed living space, *Environ. Res.,* 13, 1, 1977.

73. **Spengler, J. D. and Sexton, K.,** Indoor air pollution: a public health perspective, *Science,* 221, 9, 1983.

74. **Benson, F. B., Henderson, J. J., and Caldwell, D. E.,** Indoor-outdoor Air Pollution Relationships: A Literature Review, EPA Publ. No. AP-112, Washington, D.C., 1972.

75. **Fanger, P. O. and Valbjorn, O.,** *Indoor Climate,* Danish Building Research Institute, Copenhagen, 1979.

76. **Spengler, J. D., Ferris, B. G., Dockery, D. W., and Speizer, F. E.,** Sulfur dioxide and nitrogen dioxide levels inside and outside homes and the implications on health effects research, *Environ. Sci. Technol.,* 13, 1276, 1979.

77. **Hollowell, C. and Traynor, G.,** Combustion-generated Indoor Air Pollution, U.S. Department of Energy Publ. No. LBL-7832, Washington, D.C., April 1978.

78. **Sterling, T. and Kobayashi, D.,** Use of gas ranges for cooking and heating in urban dwellings, *J. Air Pollut. Control Assoc.,* 31, 162, 1981.

79. **Goldstein, I. F., Hartel, D., and Andrews, L.,** Indoor Exposure of Asthmatics to Nitrogen Dioxide, to be presented at Indoor Air '87, Stockholm, *Environ. Int.,* in press, 1985.

80. **Taplin, P. S. and Creer, T. L.,** A procedure for using peak expiratory flow rate data to increase predictability of asthma episodes, *J. Asthma Res.,* 16, 15, 1978.

81. **Ferris, B. G.,** Epidemiology standardization project, *Am. Rev. Resp. Dis.,* 118, 1, 1978.

82. **Korn, E. L. and Whittemore, A. S.,** Methods for Analyzing Studies of Acute Health Effects of Air Pollution, Technical Report 325, SIAM Institute for Mathematics and Society, May 1979.

83. **Korn, E. L. and Whittemore, A. S.,** Methods for analyzing studies of acute health effects of air pollution, *Biometrics,* 35, 795, 1979.

84. **Whittemore, A. S. and Korn, E. L.,** Asthma and air pollution in the Los Angeles area, *Am. J. Public Health,* 70, 687, 1980.

Chapter 10*

ASTHMA AND AIR POLLUTION IN LOS ANGELES

Roger M. Katz and Ron G. Frezieres

TABLE OF CONTENTS

* The work on which this chapter was based was a team effort involving as well, Mrs. Anne H. Coulson, Dr. Roger Detels, Dr. Sheldon Siegel, Dr. Gary S. Rachelefsky, and was supported by the California Air Resources Board and Allergy Research Foundation. The authors wish to express their gratitude to each of these and to the patients without whose cooperation the work would not have been possible.

I. THE PROBLEM

Allergists and their patients in Pasadena, California were among the first to register specific complaints concerning health effects of "smog" in the Los Angeles Basin in 1954. They complained to the local health department and air pollution control district and their complaints were picked up by the local newspapers, — it so happened — at the time of the election campaign in the months of September and October 1954. They complained that during "smoggy" weather, asthma patients were worse, and they wanted something done about it.

During the previous decade, the symptoms of "smog" had gradually worsened in the basin; they were four in number:

1.   Interference with visibility by a brownish haze.
2.   A peculiar odor, later identified as ozone, primarily.
3.   A characteristic form of plant damage.
4.   Eye and respiratory tract irritation.

Professor Arie J. Haagen-Smit at the California Institute of Technology, who was doing research on plant hormones in seedlings in greenhouses in Pasadena found that his experiments were being interfered with by the pollution, and sought to understand the mechanisms. He was able to demonstrate that the principal ingredient in the smog was ozone, and that it was formed in the atmosphere by the interaction of sunlight on a mixture of hydrocarbon vapors (mostly from gasoline and motor vehicle emissions) and oxides of nitrogen (which was emitted from both motor vehicles and other sources of combustion, such as power plants).

Ozone had been thought to be a healthful minor ingredient in natural mountain air, but when it was artificially produced during welding, the men exposed had complaints of cough and shortness of breath. Experiments with animals showed that long-term exposures led to fibrosis of the lungs, and that short-term exposures led to the lungs being filled up by fluid. It had then been recognized that ozone was a relatively toxic gas, and the levels considered to be acceptable for workmen were set at 0.1 ppm for an 8-hr work day. The levels occurring in Los Angeles were often that high, and occasionally reached 0.5 ppm for an hour.

The widespread deterioration in visibility, the rapid population growth of the area, the identification of a toxic gas in the air, and the uncertainty as to future health problems, gave to studies of air pollution health effects a sense of urgency and priority, which, after the issue was prominently discussed in the election period by candidates, led first the California State Health Department and then the U.S. Public Health Service to allocate funds and manpower to investigate health effects of what was then beginning to be called photochemical air pollution.

Two scientists were detailed by the U.S. Public Health Service to the California Health Department and were assigned to investigate the complaints of asthma aggravation in Pasadena. They received the willing cooperation of the allergists and air pollution control authorities. A panel of patients was established to maintain a diary of their health problems, the results of which could be related to air pollution measurements, meteorological factors, and pollen counts.

At first the investigators, Dr. Charles Schoettlin and Dr. Emanuel Landau were disappointed in the results *because only a minority of the asthmatic subjects seemed to be reacting to any of the pollutants measured.* They also noted that most of the asthma attacks occurred at night, whereas the peak pollution levels were in the midday to early afternoon. We now know that there is a latent period between the pollution exposure and the maximum effect and that the occurrence of asthma at night is a characteristic of asthma. Eight of their 137

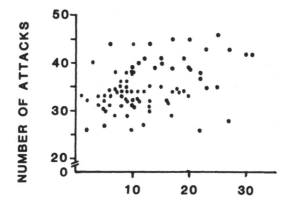

MAXIMUM HOURLY OXIDANT LEVEL (PPHM)

FIGURE 1. The correlation between the numbers of attacks of asthma of a panel and the maximal hourly oxidant levels for the previous day in Pasadena, California in the fall months of 1956. (From Schoettlin, C. E. and Landau, E., *Publ. Health Rep.*, 76, 545, 1961.)

subjects did appear to be reacting more on days when ozone levels were high, and for the whole of the panel when the days with ozone above an hourly average of 0.20 ppm for at least 1 hr were examined statistically, it was found that there were significantly more attacks on these days than on days with lower levels of ozone. (Figure 1 shows this relationship.)

After its publication and scientific review, *the Schoettlin and Landau study of asthma aggravation by oxidant (ozone) became a key piece of the scientific evidence that the California authorities used to set air quality standards and to press for the regulation of motor vehicle emissions of hydrocarbons and oxides of nitrogen.*

Two different lines of medical research branched off as a result of the stimulus provided by this research. Scientists began to recognize that a portion of the persons with asthma in the community were unusually sensitive to many of the inhaled air pollutants, and so they began to recruit volunteer asthmatic subjects to participate in inhalational studies of the health effects of specified amounts of pollutants during various grades of exercise. At the same time, further panel studies were undertaken, and rather than lumping all of the data for the panel members together, each member's data were considered separately. In this way the group of reactors to pollution could be identified and their reactions analyzed.

A large number of panels were studied in Los Angeles and using the data for each person as the basis for analysis, Alice Whittemore and Ed Korn were able to show that oxidant and to a smaller extent, suspended particulate matter were significantly associated with increase in the frequency of asthma attacks when temperature, time of year, day of week, and previous asthma experience of the subject were also included in the estimating equations.

Among the pollutants of greatest concern were the pollutants which were related to the sulfur dioxide emitted from power plants burning coal or fuel oil; this included sulfur dioxide itself and its oxidation product, particulate sulfates. To avoid such pollutants from being emitted into the Los Angeles atmosphere, the local air pollution control authorities had been authorizing only low sulfur fuels for power plants, but in the early 1970s it appeared that the supply of natural gas, which was low in sulfur was diminishing. They faced the prospect that more sulfur may be emitted from the coastal power plant complex, and so asked local medical scientists to try to determine if sulfur dioxide and sulfate aerosol exposures were aggravating asthmatics. The panel study which is the subject of this chapter was undertaken in response to this concern.

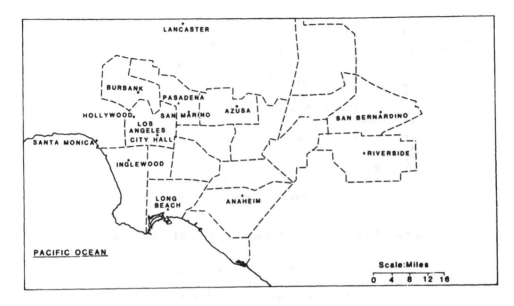

FIGURE 2. Map of Los Angeles basin with the location of the principal communities and in dotted lines the "source-receptor" areas defined by the South Coast Basin Air Quality Maintenance District.

II. THE SETTING

Los Angeles County covers approximately 9000 square miles in Southern California, much of it bordered on three sides by mountains which make the pollutants its population of nearly 7,000,000 people generate difficult to dilute. It has joined with adjacent counties to monitor its air pollution (Figure 2). The location of communities and monitoring sites is shown in Figure 2. On the basis of at least 10 years of data on oxidants, SO_2, sulfates, and total particulate matter, it was possible to identify locations which had relatively high levels of sulfates and low levels of photochemical oxidants. At one such site, Lennox, it was possible to recruit subjects for study who lived within 3 miles of the monitoring station. The area is in the southwestern portion of Los Angeles County, just east of the Los Angeles International Airport and about 5 miles inland from the Pacific Ocean. From both laboratory and field studies, it was known that sulfate levels tend to be higher on days with low wind velocity, and high oxidant levels for that site. Relatively high humidity seems to increase the transformation of SO_2 to sulfate in the atmosphere, and the closeness to the Pacific Ocean would provide for this.

The identification of this location and the feasibility of such a study were determined by a cooperative research team including scientists from the California State Department of Health, The Department of Epidemiology of the University of California's School of Public Health (UCLA), and the Medical Director of the American Lung Association of Los Angeles County.

III. THE OBJECTIVE

The objective of the study was to determine from a panel study of asthmatic subjects if some of them were reactive to naturally occurring levels of pollution with particulate sulfates.

Table 1
AGE, SEX, AND ETHNIC
COMPOSITION OF THE
ASTHMA PANEL (LENNOX)

	Age		
	9—18	19—43	44—58
Male	(17)	(4)	(2)
White	11	3	2
Black	4	1	—
Other[a]	2	—	—
Female	(2)	(6)	(3)
White	—	6	2
Black	2	—	1

[a] One each American Indian and Cuban.

IV. METHODS

A. Selection of Panel Members

All potential participants were attending an asthma clinic, and all met the criteria for the diagnosis of atopic (extrinsic) asthma established by the American Thoracic Society. Fifty-nine possibly eligible subjects were identified, all of whom were sick enough to require daily bronchodilator therapy for symptomatic relief. All were asked if they would be willing to cooperate over a long period of time in the maintenance of a diary of symptoms and medication, and perform a lung function test three times a day with a portable device. Candidates were also asked to maintain a diary for a trial period of 2 weeks. After this pretest and informed consent procedure, 34 of the subjects were empanelled and offered a stipend at the termination of their participation.

All of the selected subjects resided within 3 miles of the Lennox monitoring station, mostly downwind. There were 23 males and 11 females. Only 2 had lived in the area less than 1 year and all but 9 had lived in the area 5 years or longer. Because of the need for subject cooperation, the minimal age accepted was 9 years. The age, sex, and ethnic makeup of the panel is shown in Table 1.

B. Daily Symptom and Medication Diary

Diaries were to be filled out each morning for events occurring the preceding evening and at the end of the day for events occurring during the day. The information requested included the number of asthmatic episodes or occurrence of specific symptoms. An "asthma attack" was defined as noticeable respiratory change from normal in the opinion of the patient. The severity of each event was to be graded from 1 to 6, where 1 represented trivial or doubtful and 6 represented severe or intolerable symptoms. Such scores were requested for the specific symptoms of: wheezing, breathlessness, chest tightness, cough, sputum production, nose stuffiness, and attacks of asthma. Although the diary contained information for recording the lung function tests, it soon became clear that the subjects were apparently recording these tests inconsistently. This part of the study was therefore terminated.

Each of the daily medications was also scored on the basis of 0 to 6, with 0 indicating that the medication was not needed and 6 that hospitalization was required. The scoring guide was 1 = normal maintenance medication, 2 = addition of steroids to maintenance medication, 3 = addition of medication other than steroids, 4 = the addition of medication plus the use of steroids, 5 = emergency room or doctor visits with adrenalin or nebulization of bronchodilators, and 6 = hospital stay (including intravenous therapy).

C. Air Pollution Monitoring

Pollutant levels were monitored using the standard methods for oxides of nitrogen, ozone (oxidant), carbon monoxide, and SO_2 at the Lennox monitoring station. Oxides of nitrogen were monitored using the colorimetric (Saltzman) reagent, ozone by analyzers calibrated against UV photometric standards, carbon monoxide by nondispersive infrared spectroscopy, and SO_2 by the conductometric (H_2O_2) method. In addition 24-hr sulfate levels were monitored using the California Air and Industrial Hygiene Laboratory (of the California Department of Health Services) method #61. Twenty-four hour total suspended particulate matter was measured using the Hi-Vol Sampling method. All measurements were made for the period August 1, 1977 to April 30, 1978.

Pollens, molds, and algae were sampled on a daily basis using a roto-rod sampling device. Slides were read and reported as the numbers of a particular species of pollen, mold, or algae per cubic centimeter of air.

V. ANALYTICAL STRATEGY

Three different (but not necessarily independent) strategies were chosen. They are the examination of correlation coefficients between symptom scores, medication scores, and sulfate levels by day for each individual; the comparison of symptom and medication scores on high sulfate days compared to low sulfate days for each individual (a high sulfate day was defined as one with a 24-hr value greater than 10 $\mu g/m^3$, and a low sulfate day a value less than 5 $\mu g/m^3$), using a t test to examine the difference between means; the third strategy was to look for deviations from mean symptom and medication scores on days above and days below 10 $\mu g/m^3$ of sulfate, using a chi square test to look for the proportion of days that the individual's score was above or below their mean in association with days above or below 10 $\mu g/m^3$.

VI. RESULTS

The mean values for symptom scores during the day and during the night did not differ significantly by day of the week, nor did medication scores for the panel as a whole. In two subjects, day symptoms were greater on certain days, and in but one subject night time symptoms were greater on a specific day of the week. Therefore no day-of-week adjustment was made.

There was a weak correlation of day with night symptom scores r = 0.19.

Oxidant and sulfur dioxide were significantly correlated with daily sulfate levels (oxidant, r = 0.39; SO_2, r = 0.42). Sulfate levels were also correlated with daily minimal temperature and dewpoint, r = 0.25. Sulfate levels were negatively correlated with carbon monoxide (r = −0.23), nitric oxide (r = −0.36), and total oxides of nitrogen (r = −0.24). There were no significant associations between any of the 29 pollens identified or with any of the 30 types of molds or algae.

Table 2 shows the results of the first strategy, examination of correlation coefficients. All participants with a pollutant correlation, either positive or negative above 0.25 with symptom or medication score are tabulated. The respondents are grouped according to whether they appear to be consistently responsive to sulfate (the responders), whether they have one score significantly and positively associated with sulfate levels (partial responders), or respond to some other pollutant or respond negatively to sulfate (nonresponders).

There are thus, 8 out of the 34 panel members who show evidence of aggravation of their asthma when sulfate levels are high by at least one of the three criteria, symptom score

Table 2
POLLUTANT CORRELATIONS FOR PARTICIPANTS IN THE ASTHMA PANEL STUDY WHOSE DAY OR NIGHT SYMPTOM SCORE OR WHOSE MEDICATION SCORE CORRELATED SIGNIFICANTLY WITH ONE OR MORE POLLUTANT MEASUREMENTS (r > ± 0.25)

| Participant & category | Correlations of sulfate with | | | |
| | Symptoms | | | |
	Day	Night	Medication	Other pollutants[a]
Responders				
a	0.35	0.32	0.42	
b	0.37	0.28	0.26	
c	0.39	0.14	0.34	
Partial responders				
d	0.04	0.11	0.35	
e	0.13	0.06	0.20	NO_x, r = −0.41; CO, r = −0.40
f	0.17	0.28	0.01	
g	0.05	0.17	0.25	NO_x, r = 0.55; CO, r = 0.57
h	0.28	−0.07	0.10	
Nonresponders				
i	0.06	−0.30	0.34	
j	−0.36	−0.37	−0.04	NO_x, r = 0.22; CO, r = 0.21
k	−0.22	−0.20	−0.32	SO_2, r = −0.32; NO_x, r = −0.28
l	−0.20	−0.21	−0.11	Oxidant, r = 0.26; SO_2, r = 0.13

[a] Only selected correlations cited.

during the day or night, or medication score. An additional panelist (i) has paradoxically a negative correlation with night symptoms, but a positive correlation with medication score. Because of this inconsistency, this panelist is included with nonresponders. Each of the nonresponders responds more positively with another of the pollutants and each has a negative correlation with sulfate levels. One of the nonrespondents (k) had a negative correlation with all of the pollutants. None of the other panelists had significantly elevated correlations with any pollutant.

Two of the respondents, (b and c) also had significant correlations with total oxidants (ozone), and one (c) with SO_2.

The magnitude of these correlations was less than with sulfates.

The results for the analysis using strategy B, testing of the mean differences between scores on days with high and days with low sulfate are shown in Table 3.

These results are similar to those with the first strategy as shown in Table 2. Three of the panelists are consistently significant in the associations of symptoms and medication needs with sulfate levels. Two other are partial responders, while four have a significantly negative t test for at least one score.

The third strategy leads to results shown in Table 4. If we dichotomize (i.e., split into two groups) the days according to whether the sulfate level was above or below 10 $\mu g/m^3$, and then also look to see whether the scores are higher than the individual's mean on each day, we may test the resulting distributions with a chi square test.

We see that again the same three individuals with consistent positive associations with sulfate exposures are still showing consistent associations with the chi square criterion for day symptoms, night symptoms, and medication scores. Furthermore, the same four individuals who have some inverse association are identified, but an additional individual is

Table 3

***t*-TEST RESULTS FOR ASTHMA PANEL MEMBERS' DAY SYMPTOM, NIGHT SYMPTOM, AND MEDICATION SCORES FOR THOSE WHOSE TESTS WERE GREATER THAN 3.0 WHEN HIGH AND LOW SULFATE DAYS ARE CONTRASTED (HIGH SULFATE DAYS > 10 μg/m³; LOW SULFATE DAYS < 5 μg/m³)**

Participant & category	Day symptoms	Night symptoms	Medication
Responder			
a	5.4	5.2	3.2
b	4.8	4.4	2.7
c	2.3	(1.6)[a]	5.0
Partial responder			
d	(1.7)	2.5	3.5
h	3.3	(−0.3)	(1.1)
Nonresponder			
i	(0.4)	−4.9	3.7
j	−5.5	−4.2	(0.1)
k	−2.1	(−1.7)	−3.1
l	−2.7	−3.3	(0.0)

Note: *t*-Test results above 2 are significant at a level of $p \leq 0.05$ (less likely than 1 in 20).

[a] The values in parentheses are *not* significant, i.e., $p > 0.05$.

found. Except for subject (j) with a negative association with all pollutants, there is as before no consistent pattern of the nonresponder group. Several panelists not identified by the other criteria have either day or night symptom scores elevated significantly more frequently in association with sulfate levels over 10 μg/m³.

VII. INTERPRETATION

This pattern of a consistent response of a subset of patients we can now recognize as characteristic of asthma panels. We can conclude that small increases in sulfur oxides can lead to aggravation of asthma in a fraction of the asthmatic persons exposed.

Whittemore and Korn showed that there was about a 60% increased likelihood of an attack on a day with more than 0.3 ppm of oxidant compared to a day with oxidant of less than 0.05 ppm, and for days with oxidant above 0.15 ppm there was about a 20% increased risk. They also showed that there was a 25 % increased likelihood of an attack when the particulate matter exceeded 300 μg/m³ compared to the risk when the levels were below 50 μg/m³. Similarly our study showed evidence that the reactive group reacted to sulfate, oxidant, and SO_2. The fact that the correlations were higher for sulfate is not strong enough evidence that the effect is solely due to sulfate; indeed, it remains possible that weather conditions alone make a critical contribution to the effect.

Just as the *nonspecific reactions of asthmatic subjects* must be recognized in our interpretation, there is further a *lack of specificity in the monitoring of sulfate aerosols*. The measurement of sulfate is usually made by dissolving the soluble material from what is called a "Hi-Vol" filter. In the Hi-Vol apparatus, a motor like that of a vacuum cleaner sucks a high volume of air through a fiberglass filter which traps the particles in the air. As

Table 4
CHI SQUARE VALUES FOR TESTING WHETHER INDIVIDUALS HAVE HIGHER SYMPTOM AND MEDICATION SCORES ON DAYS WHEN THE SULFATE LEVELS ARE > 10 $\mu g/m^3$, COMPARED TO THEIR OWN MEAN SCORES, BY INDIVIDUAL AND CATEGORY (IF THE ASSOCIATION IS INVERSE, THE CHI SQUARE IS SHOWN AS NEGATIVE)

Participant & category	Day symptoms		Night symptoms		Medications	
	Chi²	p<	Chi²	p<	Chi²	p<
Responders						
a	17.1	0.00	18.8	0.00	11.4	0.00
b	3.7	0.06	4.6	0.03	14.6	0.00
c	7.9	0.01	0.3	0.66	9.5	0.00
Partial responders						
d	3.1	0.08	1.2	0.27	5.8	0.02
e	2.8	0.09	4.5	0.04	7.1	0.01
h	14.6	0.00	0.0	1.0	0.0	1.0
a′ [a]	1.7	0.19	5.6	0.02	2.2	0.14
b′	2.0	0.16	4.8	0.03	−5.1	0.02
c′	4.7	0.03	1.0	0.34	0.0	1.0
d′	0.0	0.92	0.3	0.58	4.1	0.04
e′	0.9	0.34	0.6	0.48	4.3	0.04
Nonresponders						
i	1.8	0.19	−22.8	0.00	13.5	0.00
j	−24.4	0.00	−17.3	0.00	0.0	1.0
k	2.8	0.09	2.6	0.11	−11.8	0.00
l	−4.0	0.05	−4.5	0.04	3.8	0.05
f′	0.2	0.66	0.2	0.66	−6.6	0.01

Note: Chi square tests are based on counting data. For each cell the difference between observed and expected are squared and divided by the observed; the values for each cell are then summed, and the resulting "Chi square" value is compared with tabulated results according to the number of cells.

[a] Individuals who did not appear in earlier tables have been identified as ( )′.

the particles build up in this feltlike material, the opportunity for incoming gases to react with the particulate matter already on the filter increases. Among the reactive gases are SO_2 and its oxidation product SO_3; the result of such reactions is sulfate. So when the humidity is high, the oxidant levels are high and the other particulate matter is elevated, the chance that SO_2 will be converted to sulfate on the Hi-Vol filter goes up.

There has been some interest in trying to establish an air quality standard for sulfate aerosols based on the presumption that it was the sulfate aerosol which was the main villain in the complex of pollutants of which sulfur dioxide and black suspended matter are the principal inputs. This interest was supported by laboratory tests in guinea pigs which showed that *acid sulfates* such as sulfuric acid and ammonium bisulfate (sulfuric acid in which one of the hydrogen atoms has been replaced by the ammonium ion — since it still contains a hydrogen atom it is still acid) were the most active of a large series of compounds in causing the guinea pigs to have an increase in airway resistance. Increase in airway resistance is one of the characteristic physiological features of an asthma attack.

The earliest studies of human subjects experimentally exposed to sulfuric acid or other acid sulfate aerosols did not show much effect, even in asthmatic subjects. Improvement in study design, involving use of exercise during exposures to SO_2 have demonstrated quite convincingly that physically active asthmatic subjects exposed to between 0.1 and 0.5 ppm have increases in airway resistance. When similar methods were used with sulfuric acid, the lowest effective dose was 400 $\mu g/m^3$, which is quite high when compared to the 10 $\mu g/m^3$ level which seemed to separate the days on which reactions occurred in the reactor group from the days on which it did not in our study.

At the same time this medical research was going on, the laboratory research workers were developing new methods to monitor for particulate matter according to the size of the particles. From the guinea pig experiments, we already knew that particles about 1.0 μm in diameter were much more active than particles of a size greater than about 10 μm. Since the Hi-Vol apparatus doesn't distinguish particle size, it follows that a few large particles may add a substantial amount to the weight of particulate matter, even though their size is too big to get deep into the lung on inhalation. So a new generation of monitoring equipment is being used which measures what is called "respirable particulate matter", which may be defined in various ways, but the one which is gaining preference is that it is all particulate matter of less than 10 μm. It seems likely that a new standard for respirable particulate matter will be considered, and of course it follows that we will want to look at respirable sulfate particles as well in attempting to decide what protection the regulatory process should provide for asthmatics who are pollutant reactors.

So the impact of this study is to show that asthmatic subjects in Los Angeles can have their attacks triggered by pollutants derived from sulfur compounds in fuel, possibly sulfates. It would be premature to specify a dose level at which such reactions may occur, due to the unsettled situation with respect to monitoring for particulate matter and particulate sulfates may be nonspecific. Such reactions are in addition to the effects of oxidant which were previously shown to have similar effects. Only a fraction of the asthmatic subjects show such effects, and we do not yet know how to identify the reactor group in advance, without using a provocative test. In light of the new evidence of the reactivity of physically active asthmatic subjects to experimental exposures to SO_2, it remains possible that the effect we observed is due to SO_2 exposure. Figure 3 shows the percent increase in airway resistance found for such studies according to whether the subjects breathed with a mouthpiece or through the nose and mouth (oronasal route) naturally or using a face mask. The dose figures are the product of the ppm (or $\mu \ell/\ell$) times the ventilatory rate in ℓ/min. As the figure shows, above about 20 $\mu \ell/min$ there is usually some increase in the airway resistance of a group of subjects. Possibly the most reactive ones would regularly respond to a smaller dose.

In the meantime, in Los Angeles, public warnings of expected high levels of pollutants are made so that persons with asthma can if possible restrict their activity or increase their medication level.

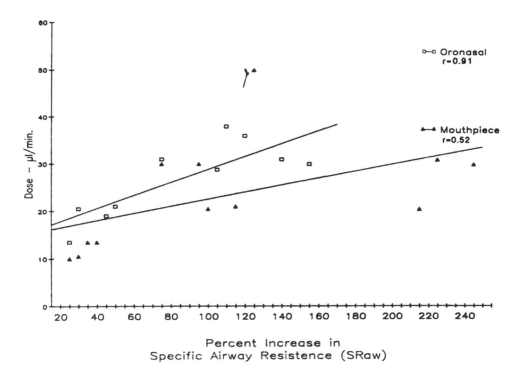

FIGURE 3. Collected data from a number of experimental studies of young exercising asthmatic subjects exposed to varying amounts of inhaled sulfur dioxide and at varying exercise rates, and either breathing through a mouthpiece or through a face mask or in an exposure chamber (oronasal). On the X axis is shown the specific airway resistance, which is the measured airway resistance, divided by the thoracic gas volume at which it is tested. (From Goldsmith, J. R., in *Air Pollution*, Stern, A. C., Ed., Academic Press, New York, 1984, chap. 6.)

Chapter 11

ODOR POLLUTION IN EUREKA, CALIFORNIA

Margaret Deane

TABLE OF CONTENTS

I. THE SETTING

For many people, especially those living near point sources of odorous emissions, air pollution is considered to be principally a problem of odor;[1] this is reflected in the complaints received by public agencies concerning unpleasant odors in the community. At the time of the studies on which this chapter is based, the subject had received little scientific attention. However, in the 1960s several studies were conducted of community reactions to odor from industries, and this information was, in some cases, used in deciding whether the odor was bothersome enough to be a basis for instituting controls.[2,3] Although some of these studies compared communities with exposure to different amounts of odor,[2-4] no studies were known to the authors which attempted to compare community reactions in areas where scientific measurements of odor had been made.

II. THE COMMUNITY

The community chosen for our studies is a northern California coastal town, Eureka, that, at the time of the initial study (1969), had about 30,000 residents. Lumbering is a major industry. Two pulp mills had been built on a peninsula to the west of the community, and residents had complained of odor that was carried eastward from the mills to the residential and business areas by prevailing offshore winds (Figure 1). Results of these studies have been published elsewhere[5-7] and the present report summarizes the most relevant findings and interprets them in the context of this book.

Three study areas were chosen to represent presumptively different exposure to odor on the basis of location with respect to the mills and the prevailing winds. These are referred to below as "exposure areas". Based on inspection of type of housing and general appearance of the neighborhoods, the areas were judged to be similar with respect to socioeconomic status. Appropriate census data were not available. About 60 households were chosen in each exposure area by systematic random sampling. One respondent was interviewed in each household, the person to be interviewed having been determined before the household was contacted.

III. OBJECTIVES

The objective of the original study conducted in 1969 was to determine whether it is possible to demonstrate the relationship between quantitative data describing odor exposure and quantitative data describing community reactions. In 1971, a follow-up study was done to assess any changes that had occurred and to obtain data on physical symptoms.

IV. METHODS

Community reactions to the odor were measured by personal interview, using a structured questionnaire. Environmental measurements were made by systematic sampling of the odor areas by a panel of trained observers using an olfactometer to sample ambient air at predetermined sites within each area.

A. Exposure Measurements

As described in an earlier publication[5] the exposure to odor in the three exposure areas was estimated by dynamic olfactometry. The measurements are based on the ratio of ambient air to odorless air at which a trained observer begins to detect malodor. This is converted to the equivalent of parts per billion (ppb) of a specific odorant by multiplying by the odor threshold of each observer, which is the ratio of a known dilution of a specific odorant to

131

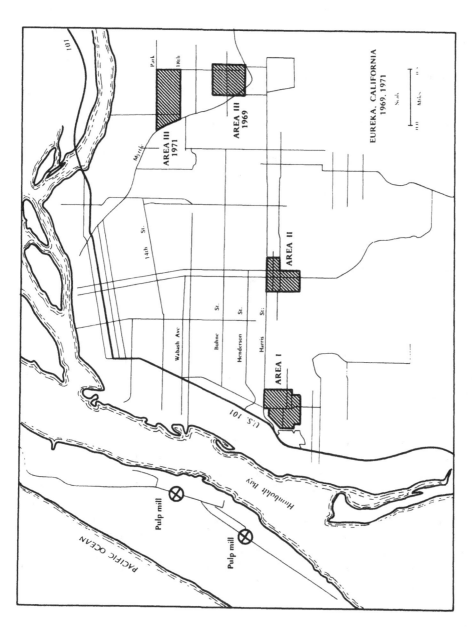

FIGURE 1. Location of pulp mills and areas studied in Eureka, California.

odorless air at which the observer begins to detect the malodor. The sensitivity of the method is a function of the observer's odor threshold and the lowest dilution of odorant to odorless air that can be measured on the olfactometer. The sampling sites, the method of sampling the ambient air, and the instrumentation and technique of using the olfactometer are described in detail elsewhere.[8,9]

B. Questionnaire and Interviewing

The interviews covered background population characteristics such as age, occupation, and family structure; general satisfaction with community conditions; attitudes toward general community problems of air pollution and noise; and frequency and intensity with which odors from the pulp mills were noticed and caused annoyance. An intensive week-long training session was held for the interviewers to minimize interviewer bias and ensure that appropriate information would be obtained from each respondent. The interview was introduced to the respondent as part of a survey on how people feel about community conditions, and no mention was made specifically of health or odor as the focus of the survey. The questionnaire was designed to give the respondent an opportunity to volunteer that pulp mill odor was regarded as a community problem before being asked specifically about it.

V. RESULTS

A. Exposure Measurements

The most important results relevant to evaluation of the annoyance reactions are the percent of total observations which indicated detection of odor and diurnal odor concentration at the 95th percentile. These confirmed the presumed odor exposure gradient of the three areas (Area I, greatest exposure; Area II, moderate exposure; Area III, least exposure) (Table 1, Figure 2).

B. Interview Survey

The results of the annoyance survey fall into four categories: (1) perception of the exposure situation, (2) annoyance reactions, (3) implications of the annoyance reactions, (4) relevance of the background variables to annoyance reactions.

1. Perception of the Exposure Situation

The frequency with which the respondents reported noticing the pulp mill odor was used as a measure of their perception of the odor. The expected gradient between exposure areas was confirmed (Table 2) with respect both to whether odor was noticed at all and to the frequency with which odor was noticed, that is, Area I showed the largest proportion of respondents noticing the odor every day or at least once a week and the smallest proportion who didn't notice it at all during the reporting period. Area III was at the other extreme, with Area II giving results between Areas I and III. A Chi square test for gradient by exposure area was done comparing those who noticed the odor "every day" or "at least once a week" with those who noticed it less frequently or reported not noticing it at all. The results were significant at the 1% level (χ^2 for trend = 49.0, 1 degree of freedom.

2. Annoyance Reactions

Those respondents who had reported noticing the odor were asked whether they were bothered by the odor very much, moderately, a little, or not at all. The data show a clear exposure area gradient with Area I having the largest percent of respondents who reported being bothered very much or moderately and Area III having the smallest percent (Table 3). The trends are less clear if one examines the individual categories of amount bothered. Combining the "very much" and "moderately bothered" categories and the "a little" and "not at all" categories results in a gradient significant at the 1% level (χ^2 for trend = 12.2,

Table 1
PERCENT MEASURABLE ODOR
DETECTIONS (ODOR FREQUENCY), 1971
BY OLFACTOMETRY

Time of day	Area I	Area II	Area III
Total 0800—1630			
Number of observations	564	846	1128
Percent with odor	37.4	14.1	5.9
800—1130			
Number of observations	256	384	512
Percent with odor	23.4	13.5	3.5
1200—1630			
Number of observations	308	462	616
Percent with odor	49.0	14.5	8.0

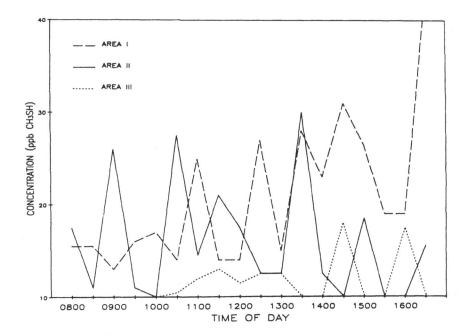

FIGURE 2. Diurnal malodor concentration at 95th percentile by area and time (as ppb CH₃SH). 1969.

1 degree of freedom). A similar expected area gradient was found for the frequency with which respondents were bothered by the pulp mill odor, but the results were based on small numbers of observations since this question was asked only of those who had already reported that they were very much or moderately bothered by the odor.

VI. FOLLOW-UP STUDY

Two years later, in 1971, the study was repeated with some modifications to determine whether community reactions had changed over the 2-year period between the two studies and to examine the relation of exposure to reported physical symptoms. Two of the same

Table 2
FREQUENCY WITH WHICH PULP MILL
ODOR WAS NOTICED, 1969
(COMMUNITY SURVEY)

	Area I	Area II	Area III
Number of respondents	52	55	51
How often odor noticed			
Every day	23.1	12.7	0.0
At least once a week	57.7	41.8	11.8
At least once a month	9.6	12.7	29.4
Less often or don't know	3.8	12.7	23.5
Not noticed at all	5.8	20.0	35.3

Table 3
AMOUNT BOTHERED BY PULP MILL
ODOR, 1969

	Area I	Area II	Area III
Number of respondents noticing odor	49	44	33
Very much or moderately bothered	53.1	38.6	27.3
A little bothered	24.5	29.5	27.3
Not at all bothered	22.4	31.8	45.5

exposure areas were used for the later study, but different households were used for the population sample. The third area was replaced with one adjacent to it because most of the households had already been sampled in the first study.

A. Exposure Measurements

The environmental measurements using dynamic olfactometry were repeated in 1971, but both the measurements of frequency with which odor was detected and the 95th percentile concentrations indicated that the area gradient of odor exposure might be more appropriately characterized by treating Area III as the moderately exposed area and Area II as the least exposed area (Table 4, Figure 3).

B. Interview Survey

When the results from the 1969 and 1971 studies were compared, the area gradient in odor perception was maintained although the magnitude was smaller (Figure 4). The same percent of individuals in areas I and II (87%) noticed the odor in 1971 as in 1969, but on the average, they noticed it less frequently. In Area III (the area with the least presumptive exposure to the odor) there was an overall decrease in the proportion noticing the odor (65 to 48%), but those who noticed it tended to notice it more frequently.

In 1971 the exposure area gradient for percent who were bothered was not as marked, and was not tested for statistical significance, although Area I still clearly showed the largest percent reporting being bothered at least moderately by the odor when the cumulative curves are examined (Figure 5). The changes in annoyance reactions between the 2 years is shown by individual category in Figure 6.

Table 4
PERCENT MEASURABLE ODOR
DETECTIONS (ODOR FREQUENCY), 1971
BY OLFACTOMETRY

Time of day	Area I	Area II	Area III
Total 0800—1630			
Number of observations	190	285	376
Percent with odor	19.5	6.0	13.3
800—1130			
Number of observations	85	123	172
Percent with odor	4.7	3.3	7.0
1200—1630			
Number of observations	105	162	204
Percent with odor	31.4	8.0	18.6

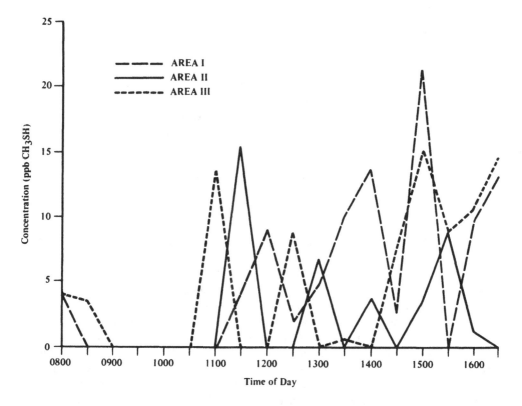

FIGURE 3.   Dirunal odor concentration at 95th percentile by area and time (as ppb CH_3SH), 1971.

C. Physical Symptoms

In addition to annoyance reactions, the 1971 interview included a list of symptoms to be read to the respondents, who were asked to report how frequently each one was experienced. In addition, respiratory symptoms from the British Medical Research Council's (MRC) standardized questionnaire on respiratory symptoms were asked.[10]

Statistically significant differences between the percents reported for men and women were noticed for headache and nervousness, and nonsignificant differences occurred in some other symptoms; therefore this part of the analysis was analyzed separately by sex. For some analyses, data were also stratified by the amount the respondent had reported being bothered

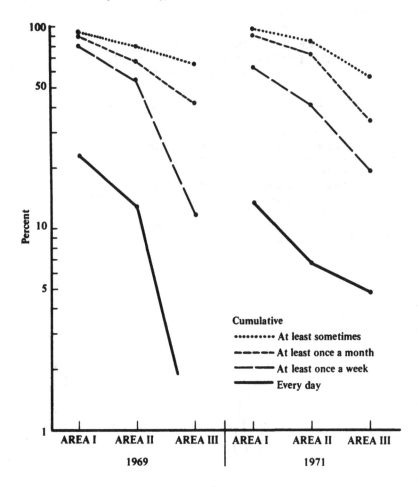

FIGURE 4.   Frequency with which pulp mill odor is noticed; 1969, 1971.

by the odor. No statistical tests proved to be significant for the hypothesized differences between areas when Area I was treated as the most exposed area and Area III was treated as the least exposed area. However, since the results of dynamic olfactometry, as well as some of the annoyance reactions suggested that a more appropriate area gradient might be represented in 1971 by using Area II to represent the least exposure and Area III to represent moderate exposure, tests were also made for this gradient. Only the percent of women reporting constipation gave a statistically significant result (Table 5). This symptom was included as a dummy variable and was not hypothesized to be related to odor exposure. When data from both sexes were combined (to obtain adequate sample sizes), and stratified by amount bothered, this result occurred only among respondents who had reported that they were bothered by the odor only a little or not at all (Table 6). Several symptoms showed significant area gradients opposite to those hypothesized from the exposure data; of particular interest is the "reverse" area gradient for eye irritation (Tables 5 and 6).

Similar tests for the hypothesized gradient between areas were made for responses to the MRC questions (Table 7). Only grade 1 or greater phlegm gave significant results, and these occurred only among women and held for both sequences of areas (Areas I, II, III and Areas I, III, II). This could be related to the higher percent of smokers among women in Area I than in Areas II and III (Table 8). The numbers of respondents were not large enough to analyze when stratified by smoking. Results of the MRC questions should not be used for

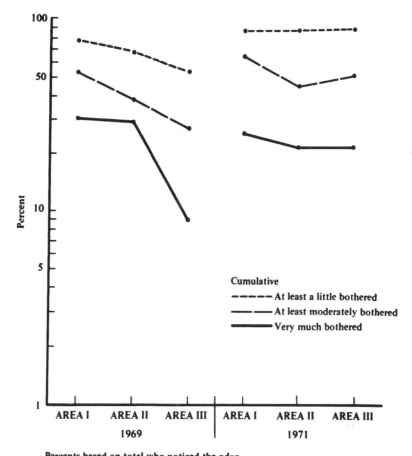

Percents based on total who noticed the odor.

FIGURE 5.   Extent to which bothered by pulp mill odor; 1969, 1971.

comparison with results of other studies using the questionnaire since modified probing rules were used, and the questions followed the section on odor annoyance, possibly biasing the responses.

For all three exposure areas combined, Chi square tests were also done for the relationship between symptoms and amount bothered by the odor. Only the percent of women reporting headache frequently or occasionally showed a significant positive relationship to amount bothered (Table 9). In spite of the lack of relationship found between this symptom and exposure area, these results support the hypothesis that headache is related physiologically to exposure to the odor, but could also be interpreted as a stress response or as the symptom most apt to trigger a "bothered" response to the questionnaire. None of the responses to the MRC questions showed a significant relationship to amount bothered by the odor.

VII. INTERPRETATION

The relationship between the exposure measurements made by dynamic olfactometry and the community responses to the odor in 1969 are shown in Figure 7. Similar exposure area gradients are observed regardless of which measurement of exposure or response is used.

The differences between 1969 and 1971 in observed exposure area gradients could be a result of several things: abatement of emissions from the mills, change in meteorological

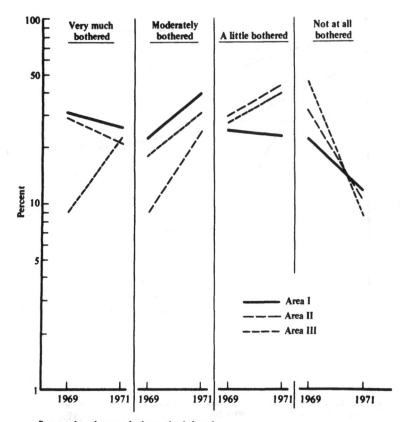

FIGURE 6.   Trends in percent bothered by pulp mill odor; 1969, 1971.

conditions, particularly wind direction and velocity, change in perception of the pulp mill odors, or change in attitude toward them.

The importance of the study lies primarily in its demonstration that measurements of perception of the odor through the use of an interview is a reliable indicator of objective measurements of odor made by trained observers using a scientific instrument, and that the interview method of measuring the degree to which a community is bothered by odor is also related.

The role of odor in causing physical symptoms is complex. If such symptoms, as measured by a health interview, are, in fact, statistically related to the frequency and intensity with which a population is exposed to odor, what does this imply? Do the odorous substances actually cause physiological change that result in the symptoms, or are the symptoms themselves an expression of an emotional reaction, such as annoyance or disgust? Add to this the possibility that the odorous substances may be an accompaniment to other airborne substances that may themselves be causing the physiological or emotional reaction.

Although laboratory studies may provide clues to the relation between exposure to odor and physical symptoms, the laboratory setting cannot duplicate the experience of man in the community with its myriad of interacting and unmeasurable forces, and it is in the community that man must live and either flourish or languish.

Table 5
PERCENT REPORTING SYMPTOMS FREQUENTLY OR OCCASIONALLY BY AREA, 1971

	Male			Female		
	Area I	Area II	Area III	Area I	Area II	Area III
Number of respondents	20	20	20	25	27	28
Symptoms						
Nervousness	35.0	30.0	30.0	72.0	70.4	78.6
Headache	25.0	25.0	30.0	52.0	37.0	50.0
Sleeplessness	20.0	30.0	40.0	40.0	29.6	32.1
Dizziness, nausea, or vomiting	5.0	10.0	0.0	12.0	7.4	10.7
Constipation	15.0	10.0	20.0	28.0[a]	7.4[a]	14.3[a]
Pain in joints	35.0	35.0	40.0	52.0	37.0	28.6
Difficulty in urinating	0.0	5.0	0.0	4.0	7.4	0.0
Sinus congestion	30.0	25.0	60.0	44.0	37.0	53.6
Eye irritation	15.0	15.0	25.0	12.0	25.9	32.1
Burning or irritation of the nose	10.0	15.0	10.0	8.0	11.1	17.9
Runny nose	20.0	30.0	25.0	28.0	25.9	39.3
Chest pains	10.0	0.0	0.0	8.0	7.4	7.1

[a] Area gradient significant at the 5% level for the sequence Area I, Area III, Area II (Chi square for trend = 4.01 with 1 degree of freedom).

Table 6
PERCENT REPORTING SYMPTOMS FREQUENTLY OR OCCASIONALLY BY AREA WITHIN CATEGORY OF AMOUNT BOTHERED BY PULP MILL ODOR, 1971

	Very much or moderately bothered by pulp mill odor			Little or not at all bothered by pulp mill odor		
	Area I	Area II	Area III	Area I	Area II	Area III
Number of respondents	28	17	12	17	28	30
Symptoms						
Nervousness	50.0	64.7	75.0	64.7	46.4	53.3
Headache	50.0	52.9	50.0	23.5	21.4	36.7
Sleeplessness	32.1	35.3	33.3	29.4	28.6	40.0
Dizziness, nausea, or vomiting	10.7	11.8	16.7	5.9	7.1	3.3
Constipation	21.4	17.6	25.0	23.5[a]	3.6[a]	16.7[a]
Pain in joints	46.4	41.2	33.3	41.2	35.7	33.3
Difficulty in urinating	0.0	11.8	0.0	5.9	3.6	0.0
Sinus congestion	42.9	35.3	50.0	29.4	28.6	63.3
Eye irritation	14.3	35.3	41.7	11.8	14.3	23.3
Burning or irritation of the nose	14.3	17.6	25.0	0.0	10.7	10.0
Runny nose	35.7	35.3	33.3	5.9	25.0	33.3
Chest pains	10.7	0.0	8.3	5.9	7.1	3.3

[a] Area gradient statistically significant at the 5% level for the sequence Area I, Area III, Area II (Chi square for trend = 3.97 with 1 degree of freedom).

Table 7
PERCENT REPORTING COUGH, PHLEGM, AND SHORTNESS OF BREATH, 1971

	Male			Female		
	Area I	Area II	Area III	Area I	Area II	Area III
Number of respondents	20	20	20	25	27	28
Symptoms						
Cough grade 1 or 2	25.0	10.0	35.0	24.0	18.5	21.4
Cough grade 2	10.0	5.0	10.0	12.0	11.1	10.7
Phlegm grade 1 or 2	15.0	10.0	15.0	32.0[a]	11.1[a]	7.1[a]
Phlegm grade 2	10.0	10.0	5.0	12.0	3.7	7.1
Persistent cough and phlegm	15.0	5.0	15.0	12.0	7.4	3.6
Shortness of breath						
Grade 2 or greater	40.0	30.0	25.0	40.0	33.3	39.3
Grade 3 or greater	0.0	5.0	0.0	4.0	11.1	10.7

[a] Area gradient significant at the 5% level for the sequences Area I, Area II, Area III and Area I, Area III, Area II (Chi square for trend = 5.85 and 4.01 respectively with 1 degree of freedom).

Table 8
DISTRIBUTION OF RESPONDENTS BY SMOKING, 1971

	Male			Female		
	Area I	Area II	Area III	Area I	Area II	Area III
Number of respondents	20	20	20	25	27	28
Smoking history						
Never smoked	10.0	20.0	25.0	24.0	55.6	39.3
Exsmokers	20.0	30.0	20.0	8.0	11.1	25.0
Present cigarette smokers	45.0	35.0	55.0	68.0	29.6	35.7
Pipe and cigar or mixed smokers	25.0	15.0	0.0	0.0	3.7	0.0

Table 9
PRECENT REPORTING SYMPTOMS FREQUENTLY OR OCCASIONALLY BY AMOUNT BOTHERED BY PULP MILL ODOR, 1971

	Male		Female	
	Very much or moderately bothered	Little or not at all bothered	Very much or moderately bothered	Little or not at all bothered
Number of respondents	20	35	37	40
Symptoms				
Nervousness	30.0	31.4	75.7	72.5
Headache	35.0	22.9	59.5[a]	32.5[a]
Sleeplessness	30.0	31.4	35.1	35.0
Dizziness, nausea, or vomiting	5.0	5.7	16.2	5.0
Constipation	15.0	17.1	24.3	10.0
Pain in joints	45.0	34.3	40.5	37.5
Difficulty in urinating	0.0	2.9	5.4	2.5
Sinus congestion	35.0	40.0	45.9	45.0
Eye irritation	25.0	14.3	27.0	20.0
Burning or irritation of the nose	15.0	8.6	18.9	7.5
Runny nose	15.0	25.7	40.5	22.5
Chest pains	10.0	0.0	5.4	10.0

[a] Difference by amount bothered is statistically significant at the 5% level (Chi square = 5.63 with 1 degree of freedom).

REACTIONS TO ODORS FROM PULP MILLS

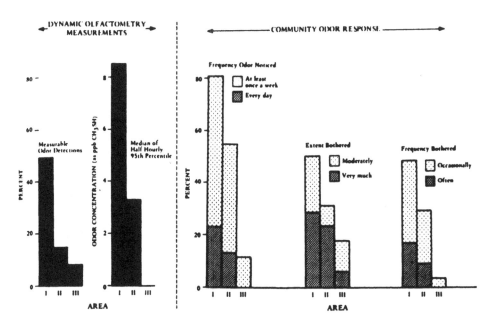

FIGURE 7.   Comparison of dynamic olfactometry measurements with community odor response, 1969.

REFERENCES

1. **Medalia, N. Z.,** Community Perception of Air Quality; an Opinion Survey in Clarkston, Washington USPHS Publ. No. 999-AP-10, Washington, D.C., 1965.
2. **Friberg, L., Jonsson, E., and Cederlöf, R.,** Studier över sanitära olägenheter av rökgaser från en sulfatcellulosafabrik (I), *Nord. Hyg. Tidskr.* 41, 41, 1960.
3. **Cederlöf, R., Friberg, L., Jonsson, E., Kaij, L., and Lindvall, T.,** Studies of annoyance connected with offensive smell from a sulphate cellulose factory, *Nord. Hyg. Tidskr.* 45, 39, 1964.
4. **Smith, W. S., Schueneman, J. J., and Zeidberg, L. D.,** Public reaction to air pollution in Nashville, Tennessee, *J. Air Pollut. Control Assoc.*, 14, 418, 1964.
5. **Jonsson, E., Deane, M., and Sanders, G.,** Community reactions to odors from pulp mills: a pilot study in Eureka, California, *Environ. Res.*, 10, 249, 1975.
6. **Deane, M., and Sanders, G.,** Health effects of exposure to community odors from pulp mills, Eureka, 1971, *Environ. Res.*, 14, 164, 1977.
7. **Deane, M. and Sanders, G.,** Trends and community annoyance reactions to odors from pulp mills, Eureka, California 1968-1971, *Environ. Res.*, 14, 232, 1977.
8. **Sanders, G. R., Umbraco, R. A., Twiss, S., and Mueller, P. K.,** The Measurement of Malodor in a Community by Dynamic Olfactometry, AIHL Report No. 86, California State Department of Public Health, Berkeley, 1970.
9. AIHL Recommended Method Number 25A, Measurement of Odor Concentration by Dynamic Olfactometry, California State Department of Health Services, Berkeley.
10. Medical Research Council's Committee on the Aetiology of Chronic Bronchitis, *Brit. Med. J.*, 11, 1655, 1960.

Chapter 12

IMPACT OF AIR POLLUTION IN THE COMMUNITY

John R. Goldsmith

One thing that Kawasaki, New Orleans, New York City, Eureka, and Los Angeles have in common is their common malady—air pollution. In each case also the complaints are related to the respiratory system's interactions with the polluted air. Increased respiratory symptoms, impairment of lung function, odor annoyance, aggravation of asthma, visits to emergency rooms for cough or difficulty in breathing are more frequent in the community afflicted by air pollution; we could say that increased frequencies of such problems are so suggestive of a community air pollution problem that they are almost diagnostic.

It is thus relatively easy to say that a community has an air pollution problem, merely by finding out if there is more frequent respiratory illness or symptoms at the times and places when pollution is increased. The development of standard interview procedures and standard ways of measuring lung function has provided tools for the diagnostician. There remain two very important and often troublesome questions which the identification of a community air pollution problem leads to.

How serious is the effect?
What needs to be done to make things better?

Probably the single most important topic to be evaluated in connection with the first question is the possible relationship between *short-term respiratory reactions and long-term health effects*. We can show that after a few minutes exposure to an odorous material in the laboratory, the awareness of the odor begins to diminish. We can study the time course of diminished lung function when subjects are exposed in the laboratory to SO_2, NO_2, and ozone, under various circumstances. We learn that the effects of SO_2 can be detected within 5 min of initiation of exposure and after exposures stop, recovery occurs well within an hour. Ozone and NO_2 have effects which are likely to develop over a period of several hours and some hyperreactivity of the respiratory tract is likely to be detectable in some subjects for several days. We have never been able to show that recovery was not complete, but this may reflect more the relatively difficult problem of long-term follow-up and relatively insensitive methods rather than the lack of such an effect. It is a hypothesis as yet unproven that frequently repeated insults to the respiratory tract by community air pollutants can lead to potentially serious chronic lung disease. It is widely believed that it is through such mechanisms that cigarette smoking has such a strong influence on the occurrence of chronic bronchitis and emphysema, one of the leading causes of death and disability.

A large scale community study looking into just such a possible effect is presented in Chapter 16.

It is for these reasons that the reactions of asthmatic subjects is felt to be so valuable despite the difficulties that Goldstein and Hartel point out in Chapter 9. Bronchial asthma is considered to be a chronic disease with periods of relative normalcy punctuated by acute respiratory distress in most cases. Persons with asthma vary greatly in the frequency and severity of their attacks, and they also vary in the long-term outcome. Whether or not air pollution is an aggravating factor, many young asthmatics will recover completely and never again be bothered by the disease. Most of these individuals have what is called "extrinsic" asthma, by which is meant that exposures to inhaled agents are thought to be the principal triggering events which bring on acute attacks. The most common agents are pollens, mold spores, and house dust, which in a sense are also pollutants. By contrast, asthma in persons over 40 is more likely to be related to disease of the heart or lungs, and in such a case would be designated as "intrinsic" asthma. The long-term outlook is not so much related to the asthma itself as to the possible underlying disease, and is therefore not quite so good. Just to make matters more confusing, there are persons with some underlying disease of the heart or lungs who are free of asthma except when they are exposed to inhalants.

Occupational exposures provide another example of a class of risks which appear to be well enough tolerated on a day to day basis but may lead to serious and disabling respiratory disease after decades of exposure. Silicosis of hard rock miners, "black lung disease" of coal miners, excess lung cancer and asbestosis of workers exposed to asbestos, a variety of chronic lung diseases among chemical workers, butchers, bakers, and candlewick makers keep us from being too complacent with respect to what are apparently reversible effects of air pollution.

There is a serious dilemma to be faced when a new type of exposure occurs. Most community air pollution problems at least seem to be new to the community which realizes that its citizens' health is at risk. The dilemma consists of either taking prompt and vigorous measures to abate the problem without ever learning if it has long-term risks or not, or alternatively, waiting for ten or twenty years to see if with improved methods and sufficient time, it can be shown that the uncontrolled pollutant has caused some excess numbers of cases of chronic disease. By that time if such effects were in fact to be occurring, they may continue to occur for several additional decades. The only really satisfactory way out of this dilemma is to prevent the short-term effects of pollution exposures, partly because they are manifestly undesirable and partly because they carry an unknown long-term risk. If this course of action keeps us from answering the question as to long-term risks, it also errs on the side of preventing harm. In some cases the prevention of short-term respiratory impacts is so technically difficult or so costly that pollution continues, and only then is it ethically attractive to look for long-term effects as is reported in Chapter 16.

In general, the only ethical course to advocate is to prevent repeated respiratory insults from air pollution.

The second critical question of course is what can be done about the offensive pollutants?

Where there is a single plant as at Eureka, the course of action may seem more obvious than when as in New Orleans, Kawasaki, and New York, there are multiple sources. Yet the emissions of pollutants from a single plant may come from many different processes, and no single one may be the culprit. Control of industrial air pollution can be a very complicated problem, and many new plants, such as the one in Eureka have episodes when the pollution control systems break down and pollutants are emitted until the engineers can correct the problem.

When there are multiple sources, the problem of knowing which one(s) need to be better controlled in order to diminish the health effects may be difficult to solve. Mapping of the location where complaints occur can help, since most respiratory exposures have complaints as a first sign. Of course there are complicated psychological and sociological processes through which effects pass before complaints are registered and these may distort the distribution which is observed in the community.

This is not the place to deal with the technical aspects of reducing specific pollutants, nor with the difference between various regulatory strategies.

However, it is often true that community studies which contrast the various effects at different places and different times can provide enough information to point to a reasonable course of action in many real but complex situations.

Part D

Chapter 13

NITRATE LEVELS IN DRINKING WATER AND METHEMOGLOBIN IN INFANTS

John R. Goldsmith

TABLE OF CONTENTS

I. THE PROBLEM

Chemicals in the drinking water may not be as great a threat to health in the 20th Century as were bacteria in the 19th Century, but we have learned to control bacterial pollution and even viral pollution. As the number of chemicals used in agriculture and industry increases, chemical pollution problems of the water supply are increasingly troublesome, especially when it is the ground water that is polluted.

The possible health threat of nitrate pollution of the ground water was only recognized in 1945 when Comley first reported that infants whose formulas were made up with contaminated ground water would get a disease called "Methemoglobinemia".[1] The well waters used were usually contaminated by barnyard waste including bacteria and nitrates as major contaminants. About 2000 cases were subsequently reported, and some of them were fatal. In 1962, in order to provide a numerical guideline for evaluation of the safety of drinking water, the U.S. Public Health Service established a drinking water standard for nitrate ion of 45 mg/ℓ as nitrate NO_3^-. Specifically it recommended that infants not consume water with more than this amount of nitrate.[2] This decision was based on a review published in 1951 reporting that no cases of methemoglobinemia had been reported in the U.S. when the level of drinking water nitrate was less than 45 mg/ℓ.[3]

There are at least three mechanisms by which nitrate in ground water can become elevated. First, there are a few locations with high levels of natural nitrate minerals which can be dissolved in ground water. This mechanism is relatively uncommon. Secondly ground water can be polluted by nitrogenous waste matter from sewage or from dairy or other livestock or poultry waste; this type of mechanism was largely responsible for the health problems recognized between 1945 and 1962. Thirdly, the use of nitrogenous fertilizers, especially in irrigated farming, carries with it the risk that the nitrates used (including such material as anhydrous ammonia which can be converted in the soil to nitrates) will seep into ground water and contaminate it. Such was the major source of nitrate pollution in the Central Valley of California; it was a contributory source to the problems of the Coastal plain of Israel, where sewage and livestock waste also contributed. In both areas adjacent wells often have widely contrasting levels of nitrates, which reflects the substantial uncertainties as to dispersion of contamination in ground waters. In addition, the level of the water table has a complex relationship between the dispersion of contaminants and the levels of such contaminants in a sample of well water.

II. METHEMOGLOBIN AND METHEMOGLOBINEMIA

Hemoglobin is the red-colored pigment of the blood which binds molecules of oxygen and allows the blood to perform its vital function of transporting oxygen to the tissues of the body. In each hemoglobin molecule there is an iron atom in the ferrous or divalent state. The oxygen molecules carried by the hemoglobin are adsorbed and do not react with this iron atom directly. If the iron atom is oxidized to the ferric (trivalent) state, the hemoglobin changes color to brown and loses its capability to carry oxygen. This brown-colored derivative of hemoglobin is called met-hemoglobin, and to a very slight extent this occurs naturally and spontaneously. In normal persons there exists an enzyme, called methemoglobin reductase, because it reduces the iron to its ferrous state and returns the hemoglobin molecule to normal, transforming brown-colored inactive methemoglobin to red-colored active hemoglobin again. In infants during the early months of life this enzyme has not yet become active and efficient, so any methemoglobin formed tends to remain in the blood cells.

When enough of the body's hemoglobin is in the form of methemoglobin, the skin appears to be dusky, rather than the usual pink color. For fair-skinned persons it usually requires about 10% of the hemoglobin to be so transformed before the color change can be detected by a trained observer. That is to say, when as much as 10% of the hemoglobin molecules are in the form of methemoglobin, it is possible to recognize that a baby has something the matter with it. Relatively simple laboratory tests may be used to see if indeed as much as 10% of the blood is in the methemoglobin form, in which case it is conventional to say the baby has "methemoglobinemia". Usually among persons in good health, about 99% of the blood is in the form of active hemoglobin, or conversely, the methemoglobin level is usually 1% or less in normal persons.

The word "methemoglobinemia" means literally methemoglobin in the blood, and in this literal sense one may speak of the normal level of methemoglobinemia of the healthy baby as less than 1%. In this chapter we shall try to restrict the use of the word to the illness or handicap associated with having a methemoglobin level above 10%. Then we shall use the percent of methemoglobin or methemoglobin level (MHb%) to indicate the situation in which the level is between 1% and 10%.

In addition to the effect of nitrate and nitrite ingestion, methemoglobin levels may be elevated because of deficiency of the enzyme methemoglobin reductase, because of the ingestion of other chemicals which oxidize hemoglobin, especially amino- or nitro-benzene compounds, and rarely because of an inherited disorder.

III. EFFECTS OF NITRATE AND NITRITE IONS ON METHEMOGLOBIN LEVELS

Somewhat paradoxically, the nitrite ion (NO_2^-) is a more active oxidizing agent than the nitrate ion (NO_3^-). So although most of the ingested indicator pollutant is nitrate, the active ion producing methemoglobin is nitrite. Nitrite can be formed in the body from nitrate by mouth and intestinal bacteria, among other ways, and this conversion can be inhibited by Vitamin C (ascorbic acid).

Several factors combine to make the young infant especially susceptible. First, the high fluid intake per unit of body weight yields a greater dose of any dissolved chemical per kilo of body weight in an infant compared to an older and larger person. Secondly, in adults, the stomach secretions are strongly acidic, and this tends to limit the types of bacteria that can survive mixture with the stomach's secretions; in the infant however, this strongly acid secretion has not yet evolved, and therefore when bacteria are ingested they may colonize the entire intestinal tract. It follows that if a young infant ingests both bacteria and substantial amounts of nitrate ion, the process of conversion to nitrite ion and its absorption may go on actively in the stomach. Finally as noted above, the enzyme methemoglobin reductase has not yet matured until about 3 months of age, and so infants of less than 3 months remain susceptible to effects of nitrate ingestion.

It is obvious that a breast fed infant is likely to have very little tap water ingestion compared to infants fed a dry or powdered milk formula, and that babies on a concentrated or evaporated milk formula would be likely to be exposed to an intermediate amount of whatever nitrate levels may be in the drinking water.

Knotek and Schmidt in Czechoslovakia[4] have shown the importance of both nitrate and bacteria in the formulas. They also showed that when similar concentrations of nitrate in drinking water were used, a dry milk formula made with a buttermilk base was less likely to lead to methemoglobinemia than when a whole milk or skim milk formula was used.

Nitrate and nitrite are also present naturally in foods such as carrots and spinach and as an additive to bacon and sausages, but these foods are not commonly eaten by infants. Blanc et al.[5] reported for example, eight cases of methemoglobinemia, half of which followed feeding of pureed carrots to babies of 4 months or less. The other four cases all had severe diarrhea. Drinking water quality did not seem to be involved.

IV. THE COMMUNITY

The study was done in two communities, Delano and McFarland, five miles apart in the Central Valley of California. Irrigated agriculture (largely cotton and grapes) is the dominant industry in the area and many of the farm workers reside in these two towns. A substantial proportion of them are of Mexican origin and speak Spanish in preference to English. A variety of wells provide the supply for these towns and some of them were known to have nitrate levels above 45 mg/ℓ from time to time. The well supplies are mixed and piped water is supplied to nearly all dwelling units from two reservoirs.

V. THE HYPOTHESIS

We wish to test the hypothesis that there is an association of methemoglobin levels with the nitrate ingestion of infants at varying ages up to 1 year. Since no cases of clinical methemoglobinemia had been reported despite the elevated levels of nitrate, we did not expect to see differences in manifest disease, but merely in the level of methemoglobin, in the under 10% range. Our target population was all infants born in the community, and we intended to examine each infant at 1 month, and at 2, 3, 4, and 6 months if possible.

VI. ACCESS TO THE POPULATION

The Department of Health of California had proposed to do such a study, but it was rejected as a budget-cutting measure, before being submitted to the legislature. Given the Public Health Service Drinking Water Standards, the Department was obliged to notify users of the high nitrate water supplies of the possible hazard of using such a water supply in infant formulas. The community, which was the center of the Farm Workers Union and had been experiencing labor disputes, was politically alert and mothers asked the city council to provide bottled water for their infants. The council asked their state senator to either find out if the bottled water was really necessary or to find funds to pay for it. The state senator happened to be a veterinary physician, well aware of the health hazards of nitrate ingestion in cattle, and asked the health department to conduct a study. Since he was also Chairman of the Senate Finance Committee the project was promptly funded.

VII. STUDY DESIGN

A Spanish-speaking nurse-epidemiologist was recruited to head the study team, and special clinics were held twice a month at hours convenient to the parents. Babies eligible were identified from hospital discharge or birth records, and families were sent appointments and contacted by a locally resident health worker as well. At each visit, nitrate and nitrite in water and formulas were determined, along with analysis for bacterial contamination. Well waters were tested weekly. At each clinic visit a 24-hr dietary and water intake history was taken, as well as an interval illness history and current symptoms were asked about. A capillary blood sample was taken and analyzed within half an hour for methemoglobin and for hemoglobin. The home of each infant was visited to obtain tap water samples, along with bacterial counts, both when taps were first opened and after flushing. Participating infants were also given a general health examination.

The study was discussed with the community's physicians, and all results were discussed with the parents as well as being sent to the baby's physician.

Since some of the infants were breast fed (even if a mother ingests nitrate-rich water, her breast milk does not have elevated levels) or were drinking liquid formula or using bottled low-mineral content waters, this group provided a reference population for comparison with those infants ingesting varying amounts of nitrate.

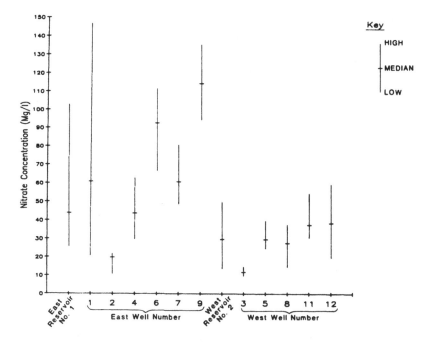

FIGURE 1. Range of nitrate concentrations in East and West Delano wells and the corresponding reservoirs.

VIII. RESULTS

Variation in the nitrate levels in the wells and reservoirs is shown in Figure 1. It can be seen that the west wells and reservoirs were generally below or near the standard, but the east-side wells, with the exception of one, usually had water levels exceeding the standard.

Over a period of 1 year, 487 examinations were completed for 256 infants; slightly more than half the infants born in the area were examined at least once. The participating infants and nonparticipating infants did not differ by ethnic origin or area of residence of their parents.

About half the infants were on concentrated formula, usually diluted with an equal volume of water. About one fourth used prepared formula and about 10% were on cow's milk to which no water was added. Nine percent used evaporated milk formulas usually diluted with equal parts of water. Less than 3% were on dry formula and 3% were breast fed. Most of the commercial formulas contain 50 mg/ℓ of Vitamin C.

About an equal number of males and females were seen, and there did not appear to be any difference in methemoglobin levels by sex.

Regardless of the water use there were substantial differences in methemoglobin by age, with virtually all the babies over 90 days having less than 2% methemoglobin as shown in Figure 2. About 25% of babies between 1 and 2 months had methemoglobin levels over 3% and about 10% of the 60 to 90-day-old babies had levels greater than 3%. As the figure shows, when the data are plotted with the cumulative frequency on a probability scale, there appears to be a bimodal distribution for the two younger age groups with the upper limb of the distribution including data for babies with greater than 4% methemoglobin. Twenty-one infants had at least one value greater than 4%, and one of these had such an elevation on two successive examinations. Accordingly, in some subsequent tables we compare the populations of children who had at least one value greater than or equal to 4% MHb with those having all their tests less. Table 1 shows the data on the distribution of MHb by age.

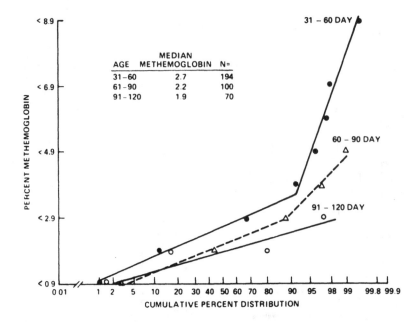

FIGURE 2. Cumulative percent distribution of the percent methemoglobin levels in Delano infants by age in days, with the median values (March 1970 to March 1971).

Twenty-one babies had one or more tests at or above 4%; of these, 16 were in the 31 to 60 day age group, three were in the next older group, and two in the next younger one. Table 2 shows the distribution of elevated MHb values according to 24-hr nitrate plus nitrite nitrogen ingestion (NNN). Overall there is a significant association between NNN ingestion and elevated MHb (Chi square = 9.7, $p < 0.001$). This association seemed limited to babies living in the areas with high nitrate in the water; as can be seen from Table 3, the nine babies with tests greater than 4% in the area served by the West Delano reservoir all had a low NNN ingestion, so the elevations were probably not due to ingested NNN, whereas 5 of the 8 babies with high MHb values in other areas, also had high NNN ingestion. Table 4 shows the data for the babies with the five highest MHb values. All of them had minor illness, either diarrhea or respiratory illness, and all had bacterial contamination of the formula. Only one, the highest, had a high NNN ingestion. Of the 21 babies with elevated MHb, 7 had respiratory illness, 5 had diarrhea, 2 had taken medication which may have caused elevated MHb and 6 had none of these problems.

No abnormalities were noted on physical examination of any of the infants, including the one with the highest MHb of 10.7%, a level at which duskiness can just begin to be detectable. One baby was hospitalized with diarrhea, and had a MHb of 8.4%, but did not have an elevated NNN ingestion.

IX. INTERPRETATION OF THE RESULTS

Babies from 1 to 2 months of age are susceptible to having elevated levels of methemoglobin, particularly if they ingest formulas made with water high in nitrate or nitrite nitrogen, if they have minor respiratory or gastrointestinal disturbances, or if their formulas are contaminated. A substantial number of such babies, regardless of these exposures, will have somewhat elevated levels of methemoglobin in the 1 to 3.9% range. The importance of minor illness in association with elevated MHb in babies from 1 to 2 months of age has not previously been recognized, and in babies who are more ill may be an important complicating factor. It is not obvious in what manner this association should be interpreted;

Table 1
METHEMOGLOBIN LEVELS OF INFANTS BY AGE IN DAYS
(DELANO, CALIFORNIA)

Methemoglobin %	Total		0—30		31—60		61—90		91+	
	No.	%	No.	%	No.	%	No.	%	No.	%
0.0—0.9	6	1.2	—	—	2	1.0	3	3.0	1	0.7
1.0—1.9	178	36.6	7	15.2	21	10.8	39	39.0	111	75.5
2.0—2.9	214	43.9	28	60.9	107	55.2	46	46.0	33	22.4
3.0—3.9	68	14.0	9	19.6	48	24.8	9	9.0	2	1.4
4.0—4.9	12	2.5	2	4.3	8	4.2	2	2.0	—	—
5.0—5.9	4	0.8	—	—	3	1.5	1	1.0	—	—
6.0—6.9	1	0.2	—	—	1	0.5	—	—	—	—
7.0—7.9	—	—	—	—	—	—	—	—	—	—
8.0—8.9	3	0.6	—	—	3	1.5	—	—	—	—
9.0+	1	0.2	—	—	1	0.5	—	—	—	—
Total	487	100.0	46	100.0	194	100.0	100	100.0	147	100.0

it may be that the formulas contamination is the cause both of the diarrhea and elevated MHb, or that the diarrhea or respiratory illness either leads to increased production of nitrite ion or other causal mechanisms for elevated MHb or that the disturbances of metabolism associated with these minor illnesses impair the reduction of the ferric iron of MHb and its conversion back to functional hemoglobin.

Younger and older babies seem less likely to have such elevated levels of MHb. The use of bottled water is capable of reversing the elevated levels, apparently, but the duration and stability of elevated levels with no change in food or water is not known.

It is plausible that elevated levels of MHb are naturally transient phenomena in babies of 1 to 2 months of age, and the time trends are not yet defined. It is reasonable to assume that some of the levels observed, would if followed at half hourly intervals, rise and some would fall.

Babies in the area served by high nitrate water supplies should be provided with bottled water to prevent the elevations of MHb; this arrangement should continue for the first three months, and is especially important for babies with diarrhea or respiratory illness.

It is not possible to interpret these data with respect to possible long-term risks, other than saying that risks, whatever they may be, should be less among babies who do not have elevated MHb during the period of early infancy.

X. POLICY IMPLICATIONS OF THESE FINDINGS

This study, by showing that there was a detectable physiological impairment associated with use of water in infant formula with elevated nitrate levels, was influential in sustaining the application of the existing drinking water standard, which previously had been interpreted as not necessarily binding. The California Health Department required that all consumers whose water supplies had levels above 45 mg/ℓ of nitrate must be notified that such water supplies should not be used for making infant formulas and that bottled water should be used instead. At least one water supply system was advised against the use of a well with high levels of nitrate.

Treatment methods for high nitrate water supplies were developed and put into use.

By publishing this report, other jurisdictions were notified that the absence of clinical

Table 2
METHEMOGLOBIN LEVELS BY AGE ACCORDING TO 24-HR NITRATE-NITRITE NITROGEN INGESTION (in mg)

Age (days)	Numbers of infants	24-hr nitrate-nitrite nitrogen[a]					
		< 5 mg/day		5—9.99 mg/day		10.0—20.0 mg/day	
		MHb<4%	MHb4% +	MHb<4%	MHb4%	MHb<4%	MHb5% +
<30	45	38	2	5	0	—	—
31—60	194	156	11	19	6	2	0
61—90	102	87	1	11	2	1	0
91 +	144	129	0	12	0	3	0
Subtotal		410	14	47	8	6	0
Total	485	424		55		6	
% with MHb 4% +	4.52	3.3		14.5		0.0	

[a] 5 mg/ℓ of nitrate-nitrite nitrogen = 22.15 mg/ℓ NO_2; 10 mg/ℓ of nitrate-nitrite nitrogen = 44.3 mg/ℓ NO_2.

illness due to elevated methemoglobin (methemoglobinemia) was not a sufficient basis for accepting the safety of high nitrate water supplies.

One very practical aspect of formula preparation came to light during the field work for the study. With all the attention given in the past to bacterial aspects of water quality, and the repeated advice to boil water to sterilize it, some of the mothers in our survey rather plausibly assumed that if there were a problem with the safety of the water, that the more it was boiled, the safer it was! In the case of the problem being a mineral contaminant, exactly the opposite is true. Boiling for up to 10 min had little effect on measured values of NNN, but from 10 to 15 minutes of boiling began to be associated with increasing NNN. In addition, in a single laboratory experiment, we observed that nitrite was produced in greater amounts when boiling was in aluminum than when it was boiled in a glass container.

In our study we did not make systemic observations as to the possible preventive effects of Vitamin C. (But see next section.)

XI. THE RELATIONSHIP OF THESE FINDINGS TO COMPARABLE AND SUBSEQUENT STUDIES

At the time of this study, a similar problem was being investigated in Israel, although with somewhat different methods. Both studies were reported at the same session of the American Public Health Association. Shuval and Gruener[7] studied 1702 infants living in communities on the Israeli coastal plain, with mean well water nitrate of 70 mg/ℓ, (the exposed group) and compared their MHb levels with those of 758 infants from the Jerusalem area with mean water supply nitrate of 5 mg/ℓ. The principal results are shown in Table 5, from which there appears to be very small differences between the exposed group (high area) and the low. Differences are somewhat greater when the babies are divided according to whether dry milk formula (powdered milk) is used, Table 6. Slightly higher MHb values are found for babies from 1 to 90 days living in the high area if there was no added Vitamin C (represented by citrus or tomato juice) in the diet; substantially higher MHb values were found for babies 1 to 90 days of age in the presence of diarrhea.

Their method for measuring MHb involved chilling the samples, after centrifugation and sending them to a central laboratory, which may account for the generally lower levels of MHb in this study, compared to the Delano study. Except for finding that the MHb is

Table 3
NITRATE-NITRITE NITROGEN INGESTION AND METHEMOGLOBIN LEVELS BY LOCATION FOR TOTAL INFANTS

| | | 24-Hr nitrogen intake | | | |
| | | <4.99 mg Mhgb level | | 5.00+ mg Mhgb level | |
Age	No. Infants examined	<4%	≥4%	<4%	≥4%
East Delano					
<30 days	14	12	0	2	0
31—60 days	63	54	1	6	2
61—90 days	40	34	1	4	1
>90 days	47	46	0	1	0
Subtotals	164	146	2	13	3
Totals		148		16	
Per cent with 4% Mhgb or more[a]		1.4%		18.8%	
West Delano					
<30 days	15	14	1	0	0
31—60 days	76	58	8	10	0
61—90 days	36	31	0	5	0
>90 days	56	50	0	6	0
Subtotals	183	153	9	21	0
Totals		162		21	
Per cent with 4% Mhgb or more[a]		5.55%		0.00%	
Other Locations					
<30 days	16	12	1	3	0
31—60 days	55	44	2	5	4
61—90 days	26	22	0	3	1
>90 days	42	34	0	8	0
Subtotals	139	112	3	19	5
Totals		115		24	
Per cent with 4% Mhgb or more[a]		2.6%		20.8%	

[a]   4.52% of the total infant population have Mhgb above 4%.

5 mg/ℓ nitrogen 22.15 mg/ℓ NO_3
10 mg/ℓ nitrogen 44.3 mg/ℓ NO_3

significantly higher during the first 60 days of life, none of the other differences were said to be statistically significant. Measurements of NNN for each infant was not done. Shuval and Gruener undertook some useful studies on pregnant rats given 2.5 to 50 mg/kg of sodium nitrite orally or by injection. A characteristic result shows that nitrite passes the placental barrier with a 20 to 40 min lag, and it appears that MHb once formed lasts about $^1/_2$ hr in both the mother and fetus. There appeared to be a transplacental threshold for a dose of 2.5 mg/kg. Pregnant rats seemed to have a greater elevation of MHb per dose of sodium nitrite than did nonpregnant ones.

Table 4
CHARACTERISTICS OF THE INFANTS WITH THE FIVE HIGHEST METHEMOGLOBIN VALUES, AND THE TYPE OF FLUID INGESTION

Percent MHb	Birth weight (lb, oz)	Illness	Formula type[a]	Contamination	Added water (oz)	NNN (mg/day)
10.7	9—6.5	D[b]	Evap. Dext.	+	8	6.31
8.9	7—6	D	Conc. 1:1	+	0	0.04
8.9	7—6	D	Prep.	+	0	0.17
8.4	7—3.5	D	Dry	+	3.5	1.16
6.1	7—5	R	Evap.	+	0	2.89

[a]   The abbreviations used for formula type are Evap., evaporated milk. Dext., added dextrose; Conc.1:1, concentrated milk diluted with equal parts of water; Prep., prepared formula; and Dry, the formula is made up from dry milk powder.

[b]   D stands for diarrhea and R stands for respiratory illness. The first infant had had loose stools for 30 days.

Table 5
MEAN METHEMOGLOBIN LEVELS AMONG INFANTS BY AGE IN DAYS IN AREAS WITH LOW AND WITH HIGH MEAN NITRATE IN DRINKING WATER (ISRAEL)[7]

Area	Mean No$_3$ in water (mg/ℓ)	Total population			1—60		61—90		91+	
		No.	Mean MHb%	SD	No.	Mean MHb%	No.	Mean MHb%	No.	Mean MHb%
Low	5	758[a]	1.11	0.72	96	1.30	75	1.24	556	0.97
High	50—90	1702	1.01	0.72	71	1.38	188	1.14	1426	0.99
Total		2473	1.04	0.72	167	1.33	263	1.17	1982	0.98

[a]   Includes 31 tests done for infants for whom age was not recorded in the low area and 17 for whom age was not recorded in the high area.

Table 6
METHEMOGLOBIN LEVELS IN INFANTS DRINKING POWDERED (DRY) MILK FORMULAS AND IN THOSE INFANTS USING ALL OTHER FORMS OF MILK BY AREAS WITH HIGH AND LOW NITRATE IN DRINKING WATER (ISRAEL)

Area	Powdered milk formulas		Other forms of milk	
	No.	Mean MHb%	No.	Mean MHb%
Low	111	0.98	664	1.14
High	36	1.17	1666	1.00
Total	147	1.01	2310	1.04

In their discussion, Shuval and Gruener say that…"no apparent public health problem associated with infant methemoglobinemia was detected in the study area, despite the fact that most of the wells were supplying water with nitrate concentrations above that generally recommended by public health authorities." They believed this to be due to the unusually frequent use of Vitamin C supplementation. Later, when the analysis was repeated with babies of 31 to 60 days alone, significant differences were found.

XII. THE CANCER CONNECTION

There is a hypothetical relationship between ingested nitrate and cancer risks, based on the possible contribution of such nitrate to the formation of a potent class of carcinogenic agents in the body. The class of agents are called nitrosamines and in laboratory animals they are capable of causing cancer in very small doses. No cancer in man has been as yet related to exposures to nitrosamines. It is known that they may be formed in the stomach by the nitrosation of amines which are commonly formed during the digestion of a number of foods. Ingested nitrate and nitrite can provide the nitroso ion. There is some evidence that stomach cancer in man is related to a diet lacking fresh fruit and vegetables, and it is known that Vitamin C can inhibit the nitrosation reaction. Fresh fruit and vegetables contain an abundance of Vitamin C. Stomach cancer is very frequent in parts of Chile and Japan, although in the U.S. and Western Europe the rates have been falling steadily in the last 30 years. Some current research is trying to see if there is a link between stomach cancer rates in Chile and ingestion of nitrate in water and food. One effort to see if there was a relationship of nitrate in water and the distribution of stomach cancer in Israel failed to show any effects, but it is thought that the steps in causing a change from normal to potentially cancerous cells in the lining of the stomach may take many years. Most of the Israeli residents of areas with elevated nitrate in drinking water have had relatively brief exposures.

Of course it would be unethical to administer nitrosamines to human subjects in order to see if they developed cancer. So the possibility for further information on this possible relationship depends on further epidemiological studies and over a long period of time. In the meantime, the possible cancer connection, which is only a hypothetical one, is an additional reason for not allowing easily avoidable excess ingestion of nitrate in drinking water, or for that matter in food.

REFERENCES

1. **Comley, H. R.,** Cyanosis in infants caused by nitrates in well water, *JAMA*, 129, 112, 1956.
2. Public Health Service Drinking Water Standards, Revised 1962, U.S. Department of Health, Education and Welfare, Public Health Service, Washington D.C.
3. **Walton, B.,** Survey of the literature relating to infant methemoglobinemia due to nitrate contaminated water, *Am. J. Publ. Health*, 41, 986, 1951.
4. **Knotek, Z. and Schmidt, P.,** Pathogenesis, incidence and possibilities of preventing alimentary nitrate methemoglobinemia in infants, *Pediatrics*, 34, 78, 1964.
5. **Blanc, J.-P., Teyssier, G., Geyssant, A., and Lauras, B.,** Les methemoglobinemies au cours des diarrhees aigues du nourrisson, *Pediatrie*, 38, 87, 1983.
6. **Shearer, L. A., Goldsmith, J. R., Young, C., Kearns, O. A., and Tamplin, B. R.,** Methemoglobin levels in infants in an area with a high nitrate water supply, *Am. J. Publ. Health*, 62, 1174, 1972.
7. **Shuval, H. I. and Gruener, N.,** Epidemiological and toxicological aspects of nitrates and nitrites in the environment, *Am. J. Publ. Health*, 62, 1045, 1972.

Chapter 14

CARBON MONOXIDE EXPOSURES AND SURVIVAL FROM HEART ATTACKS

John R. Goldsmith

TABLE OF CONTENTS

I. THE POLLUTANT

Carbon monoxide is a colorless, odorless gas, produced when a fire or other type of combustion proceeds with insufficient oxygen. Prior to introducing motor vehicle exhaust controls, it used to occur in concentrations of about 1% in gases escaping in motor vehicle exhaust. It is a very important contributor to the hazards of cigarette smoke, and many types of heating and cooking fires may produce carbon monoxide contamination indoors, especially during cold weather when the doors and windows are tightly closed.[1]

Carbon monoxide (CO is the chemical formula with one atom of carbon and one of oxygen) produces most of its health effects because the gas combines tightly with the hemoglobin in the body's red blood cells. So tightly does it hold on the hemoglobin that it keeps that hemoglobin from carrying oxygen, which is what we depend on it to do. By far its most important effect is therefore the interference it causes in the oxygen transport function of the blood. One can estimate the severity of recent CO exposure of a person by measuring the proportion of his hemoglobin which is bound to carbon monoxide. The combination is called carboxyhemoglobin, and abbreviated COHb. A tiny amount of carbon monoxide is produced by the normal biochemical processes of the body; this usually results in 0.3 to 0.5% of COHb. A pack a day smoker will usually have about 5% COHb, and heavier smokers who inhale may have levels three to four times as great.

Experimental studies have shown that both the uptake and the excretion of CO occurs through the lungs and airway, and if one enters a chamber in which the concentration of CO is kept at a constant level, say 100 ppm, it will take a resting man between 3 and 4 hr to reach half of the level of COHb which he will have if he stays there indefinitely. The uptake will be quicker, if the man exercises; if he is small in stature (and blood volume) the uptake will also be quicker because he doesn't have so much blood in the body to react with the gas. From such experiments, the staff and consultants to the California Department of Health had concluded that exposure to around 30 ppm of carbon monoxide over an 8-hr period might lead most of the exposed population to have about 5% COHb.[2] As this amount of handicap was felt to be a risk to health that should be prevented, the Department set as its initial air quality standard a level of 30 ppm averaged over 8 hr. (It has subsequently been set at a lower figure.) As part of the argument for the standard it was predicted that if such a level occurred frequently, it could interfere with the survival of patients who had cardiovascular disease, particularly those with recent myocardial infarctions. No data were available to test such a prediction, which was based only on the presumption that there was a period during the course of blocking of the circulation in the coronary blood vessels in which the continued functioning of the pumping action of the heart could be critically handicapped by a small change in the oxygen carrying function of the blood. At the time the first standards were set in 1959, it was known that levels of 30 ppm were occasionally being exceeded in Los Angeles due almost entirely to motor vehicle emissions.

The hypothesis — During times and at locations in Los Angeles Basin when the pollution from carbon monoxide is great, the likelihood of survival of patients with recent myocardial infarctions (heart attack) will be less.

II. THE POPULATION AND THE COMMUNITY

It happened that the Los Angeles County Medical Association had urged the State Health Department to examine the frequency of admissions to hospitals for respiratory and cardiac conditions in order to see if these were higher on days of high pollution. As a result, the Department contacted 56 hospitals, having a bed capacity of 12,554, to ask if they would abstract from their record rooms data on the occurrence of such admissions during the year 1958. Of these, 35 hospitals including 10,170 beds agreed to participate. The Los Angeles

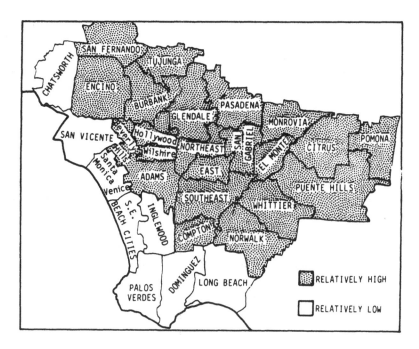

FIGURE 1. Map of Los Angeles Basin showing communities in which carbon monoxide levels were relatively high and those with lower levels (based on the 8 ppm isopleth for 1955). (From Cohen, S. I., Deane, M., and Goldsmith, J. R., *Arch. Environ. Health*, 19, 510, 1969. With permission.)

County Hospital with 2,876 beds was represented by a sample of every 20th admission. The data were tabulated by discharge diagnosis, date of admission, date of discharge, age, sex, whether the patient was discharged alive or dead, and other information.[3]

For the original hypothesis to be tested, as to whether there were increases in hospitalization for diseases of the heart or lung during periods of high pollution by oxidant, the data did not give any clear answer. With increasing interest in data to support or to reject the hypothesis on which the carbon monoxide standard was based, we decided to conduct a test of the hypothesis stated above.

Our population therefore consisted of 3080 persons admitted to 36 hospitals during the calendar year 1958, of whom 701 (26%) died in the hospital. All were diagnosed as having acute myocardial infarction.

III. THE COMMUNITY'S EXPOSURE

Five air pollution monitoring stations were in operation during the year. Since we were interested not in short-term or hourly fluctuations, we decided to take the daily average of each station as the indicator for amount of pollution present during that day. Two different combinations of such data were made. First, the average of the data for the five stations for a given day was obtained and used to represent the severity of the pollution for that day, the "daily basin average". Then, the yearly average for each station was obtained and on the basis of the areas which each monitoring station was thought to represent, the basin was divided into two areas, a relatively high pollution area, and a relatively low pollution area, according to whether it was thought that the yearly average was above or below a mean of 8 ppm. This resulted in a map (Figure 1) dividing the basin into a low pollution, generally coastal area, and a higher pollution, inland area. We recognized that the average income of the coastal area was greater, and that in general the population and possibly the types of medical care available were not identical in the two areas. Since a sizeable number of deaths from myocardial infarction occur so quickly that the patient never reaches the hospital alive, we realized that we were not able in this study to examine the total population of deaths from myocardial infarction. Furthermore, not all myocardial infarctions were diagnosed and hospitalized, so we were not examining all of the myocardial infarctions which occurred among the residents in the area.

IV. VALIDATION OF THE DATA

Apart from recognizing the possible biases in our data, we did not know of any practical way of validating hospital records for 1958 when we started this research in 1967. There was no reason to suspect any bias on the part of medical record librarians, since they were unaware of the hypothesis at the time they abstracted the records. Calibration procedures were used systematically by the Los Angeles Air Pollution Control District. Furthermore, the separations into high and low areas conformed generally with what was known of the meteorology, topography, and distribution of motor vehicle emissions in the basin. High days for carbon monoxide were generally days with low level atmospheric inversions, low winds, and hence poor dispersal of pollutants emitted from motor vehicles.

V. ANALYTICAL STRATEGY

We knew in advance that there were day-of-week effects on both carbon monoxide emissions and possibly on admissions for heart attacks. Because the selection of persons to be admitted and the staffing of hospitals to deal with emergencies which arrive can be affected by day of week, we assumed from the outset that day-of week might be a confounding factor. The reason for day-of-week effects on carbon monoxide emissions is that motor vehicle usage is different on weekends than on weekdays. Therefore at the very least we will have to examine the data for day-of-week variation, and we shall probably have to adjust for it. Furthermore it happens that in general, environmental and health conditions on successive days are not truly independent; that is today's weather and pollutant levels are more likely to resemble those of yesterday or tomorrow than they are to resemble a random day at the same time of year. This is known technically as autocorrelation in time series data. If we decide to treat the data for each day of the week separately, that is analyze all Mondays together, then all Tuesdays, etc., the problem of autocorrelation in time series is much less. Analyzing each day of the week separately then solves (or minimizes) the effects of two problems, that of day of week and autocorrelation, so that is what we shall do. We also know that there is a tendency for carbon monoxide levels to be higher in winter since dispersion conditions in Los Angeles are less favorable in the winter. Since respiratory conditions often complicate cardiac conditions and respiratory conditions are more prevalent in the winter, we might find that winter would be a period of more serious or more frequently complicated heart attacks. We must therefore look for a time of year effect in the occurrence and in the fatalities from heart attacks, as well as in the level of exposure to carbon monoxide. Some investigators have attempted to deal with this possible source of bias by looking at the deviation from a seasonally adjusted value or by looking at deviations from a moving average. When this approach is used, there always exists the possibility that some of the source of seasonal variation in morbidity or mortality may in fact be contributed by the pollution exposure, so such an adjustment would make it harder to detect such an effect. We therefore chose a different strategy. We assume that time of year effects would be likely to affect phenomena in both the low and high pollution area in the same way. If, therefore, we compare the experiences of these two areas by week of the year, and there is an effect of carbon monoxide independent of time of year effects, then the high pollution area would have relatively higher death rates during the high pollution periods, but not (or not to the same extent) during periods of relatively low pollution.

We shall be comparing case fatality rates, that is, the number of persons with myocardial infarction who die compared to the number at risk in the hospital. It is possible to make two different estimates of case-fatality rates for a given day's exposure. We can estimate the case-fatality rate for all the persons in the hospitals on a day with a given level of carbon monoxide. Or alternatively we can study the data for case fatality for the entire hospitalization period with respect to the level of carbon monoxide on the day of admission. These two estimates are designated as the "man-days-at-risk case fatality rate" and the "admission case fatality rate". Since our original hypothesis was that exposure to high levels of carbon monoxide augmented the risk for someone for whom oxygen delivery was critical, and the most critical period for a patient would be the time of initial attack, effects of carbon monoxide on the "admissions" fatality rate would be expected to be greater than on the "man-days-at-risk" rate. Of course the latter rate will be lower because each patient can contribute one day to the denominator for every day in the hospital, but the number of deaths, the numerator, is a fixed number and identical in both rate calculations.

VI. RESULTS

The data by day of week are shown in Table 1 and the time of year pattern in Figure 2. No time of year effect is obvious in the number of admissions for myocardial infarction, nor in the number of deaths, nor in the "man-days" case fatality rate by inspection. A distinct yearly trend is seen in the carbon monoxide mean level, being higher in winter. Although the "admission" case fatality rate is more variable, it does appear that the rate is higher in winter. From Table 1 we see that the carbon monoxide levels do tend to be lower on weekends than on weekdays, and the mean number of admissions also tend to be lower on weekends. When we wish to study correlations, we use the logarithm of the carbon monoxide, since the data tend to be log-normally distributed. The arc sine transformation is used for the case fatality rates, in order to better approximate a normal distribution, and to take into account the variance of a ratio. We see that for most of the days, the admissions are negatively correlated with the carbon monoxide level (although none of the correlations reach the 5% significance level). The correlation between the admission case-fatality rate and carbon monoxide level is positive for all days but one, and for two of them it is significantly so. By contrast, none of the "man-days" rates are significantly associated with carbon monoxide levels.

In Table 2, the data for high and low pollution areas are shown separately. Now it is clear that there is a positive association between the admission case fatality rate and basin average carbon monoxide mean for the high area, but not for the low area. In the high area even some of the "man-days" rates are correlated with carbon monoxide levels to a significant degree.

Table 3 shows the data by week of the year and by area. Also shown by a plus or a minus is whether the case fatality rate is greater (+) in the high pollution area or in the low pollution area. Since the overall case-fatality rates are higher in the high pollution area than in the low pollution one, we are not surprised to find that the plus signs predominate, being 35 out of 52 possible weeks. There also seems to be some seasonal pattern, with the bulk of the minus (−) falling in the Spring of the year, and the other three seasons having a predominance of weeks in which the more polluted area had the higher case-fatality rate.

However when the weeks of the year are ranked by their mean CO levels we find that there is a clear tendency for the high case fatality rates to occur during weeks with high CO levels. In fact, in 12 out of 13 of the weeks in the highest quartile, the more polluted area had the greater case-fatality rate. This result would be expected as a result of random processes

Table 1
RELATIONSHIP BETWEEN MEAN NUMBER OF HOSPITAL ADMISSIONS FOR MYOCARDIAL INFARCTION (MI), CASE FATALITY RATE, AND MEAN AMBIENT CARBON MONOXIDE LEVELS FOR LOS ANGELES BASIN HOSPITALS BY DAY OF WEEK, 1958[3]

Days	CO level	Admissions (MI)	Correlation (ADM vs. CO)	Case fatality rate (CFR)	Correlations CFR vs. CO
Sunday	6.53	6.94	−0.251	28.3	0.049
Monday	7.71	9.54	−0.234	25.4	0.019
Tuesday	7.45	9.41	−0.071	23.5	0.134
Wednesday	7.53	9.13	0.150	24.2	0.273[a]
Thursday	7.71	8.08	−0.019	26.9	−0.030
Friday	7.90	8.56	0.121	23.6	0.262
Saturday	7.07	7.58	−0.040	32.0	0.309[a]
Weekdays	7.66	8.94	−0.120	24.7	0.130[a]
Weekends	6.80	7.26	−0.022	30.2	0.177
All days	7.41	8.46	0.002	26.0	0.114[a]

Note: Case fatality rates are admission case fatalities, per 100 admissions, CO levels in ppm.

[a] Significantly above 0.00 with a $p < 0.05$. The correlations with the CFR are based on the arc sin transformation of the CFR.

less than 1 time in 20, and it is therefore considered statistically significant. The results for the four quartiles are shown in Figure 3.

VII. CONCLUSIONS AND INTERPRETATIONS

From this community study we conclude that two findings indicate that in Los Angeles in 1958 in times and places of high exposures to carbon monoxide there is an increase in the case fatality rate for myocardial infarction (heart attacks). These two findings are the positive correlations between the case fatality rate in the high pollution area but not in the low pollution area by day of week (Table 2) and the high proportion of weeks with high carbon monoxide levels for which the high pollution area had a greater case fatality rate than did the low pollution area (Table 3 and Figure 3).

We recognize that risk factors for heart attacks, the likelihood that a person with a heart attack would reach a hospital alive, and health care may differ between these two areas. Accordingly, if we only showed that the case fatality rate was greater in the high pollution area (which it was) we would not be able to distinguish whether pollution or these other factors were responsible. Similarly, there is a marked seasonal pattern in carbon monoxide levels and a corresponding but less striking increase in the case fatality rate during the winter. Therefore, if we only showed that the case fatality rate was higher during the weeks in which carbon monoxide was higher, we should have been unable to distinguish whether it was not the weather in winter which was responsible rather than the exposures to carbon monoxide.

In neither the former type of analysis by location, a spatial association, nor the latter type of analysis over time, a temporal association, could we have drawn a conclusion that was specific for possible effects of carbon monoxide. The alternative, which did allow us to draw such a conclusion was the result of what is called a "temporo-spatial" analysis. In such an analysis we tested whether the case fatality rate was increased at the time (temporo- and place (-spatial) of high exposures to carbon monoxide. For such a result to have been

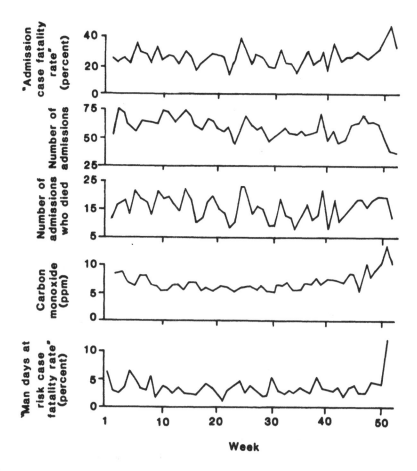

FIGURE 2. Average weekly values for numbers of persons admitted to the sampled hospitals with acute myocardial infarction, the number who died prior to discharge, and the mean weekly carbon monoxide values. The upper panel shows the proportion of those admitted who died (case fatality rate) — "the admission case fatality rate". The lower panel shows the case fatality rate pased on the proportion of persons in the hospital on the relevant day "man days at risk case fatality rate". (From Cohen, S. I., Deane, M., and Goldsmith, J. R., *Arch. Environ. Health*, 19, 510, 1969. With permission.)

due to, say health care or risk factors for heart attacks, it would have been necessary to hypothesize that these variables had their greatest effects during the time of year at which carbon monoxide was at its highest, and such an assumption is unlikely.

We were able to test for some area differences from census data and from what was recorded about the persons admitted with heart attacks. The census data for example tell us that a greater proportion of the population is over 65 years old in the high pollution area than in the low pollution area. But we did not find any differences in the ages of the men and women hospitalized with heart attacks. The ratio of men to women was 1.8 to 1 in the high area patients, compared to 2.1 to 1 among the low area patients. Slightly more of the patients from the high area had hospital insurance than those from the low area. Such differences can hardly account for temporo-spatial differences in case-fatality rate that we observed.

The findings therefore suggest but do not prove that carbon monoxide is associated with high case-fatality rates for acute myocardial infarction. These data were obtained at a time when intensive coronary care units were not widely available. Since 1958, case-fatality rates for myocardial infarction have decreased and it is possible that the widespread availability of ICCU (intensive coronary care unit) services has contributed. It is well known that if a person exposed to carbon monoxide is given oxygen to breathe, the CO is rapidly excreted

Table 2
MEAN CASE FATALITY RATE FOR MI BY DAY OF WEEK AND BY AREA AND THE CORRELATION WITH BASIN AVERAGE CARBON MONOXIDE LEVELS, LOS ANGELES, 1958[3]

	High area		Low area	
Days	CFR	Correlation CFR vs. CO	CFR	Correlation CFR vs. CO
Sunday	29.3	0.070	22.8	0.002
Monday	26.3	0.057	11.7	0.081
Tuesday	24.5	0.057	21.8	0.206
Wednesday	24.7	0.192	24.1	0.228
Thursday	29.4	0.164	18.4	−0.131
Friday	24.3	0.346[a]	23.5	−0.133
Saturday	34.0	0.482[b]	21.2	−0.193
Weekdays	25.8	0.161[b]	19.1	0.050
Weekends	31.7	0.280[b]	22.0	−0.112
All days	27.3	0.161[b]	19.1	−0.003

Note: All correlations based on arc sin transformation of CFR; CFR per 100 admissions.

[a] $p < 0.05$
[b] $p < 0.01$

from the body, and the prompt use of oxygen for persons having or suspected of having heart attacks is now widespread, and may counteract the risks of CO exposures to some extent.

VIII. IMPACT OF THE STUDY ON POLICY DECISIONS

As the first positive report of epidemiological evidence of unfavorable effects of low level carbon monoxide exposures on a sensitive group of humans, the study had a substantial impact on a series of reviews of air pollution control policies with respect to carbon monoxide during the years following its publication. Both the California Department of Health and California Air Resources Board after review set a lower Air Quality Standard for CO in 1969[4] and the U.S. Environmental Protection Agency (EPA)* did likewise in the following year.[5] The principal review was that of the U.S. National Academy of Sciences-National Research Council's Committee on the Biological Effects of Air Pollutants which published a report in 1969 on "Effects of Chronic Exposure to Low Levels of Carbon Monoxide on Human Health, Behavior and Performance".[6]

It is impossible to isolate the impact of this community study on policy decisions which followed it, since the report of this study led to a series of clinical investigations and additional epidemiological studies, which together tended to support the general conclusion that impairment of oxygen delivery by carbon monoxide would be especially harmful to persons with pre-existing cardiovascular disease. What is clear is that the conclusions of this study pointed to an important area of research which at least three groups of clinical investigators

---

* At that time designated National Pollution Control Administration

Table 3

COMPARISON OF MYOCARDIAL INFARCTION CASE FATALITY RATE BY WEEK BY HIGH AND LOW POLLUTION AREA AND WEEKLY BASIN AVERAGE CARBON MONOXIDE LEVELS (in ppm), LOS ANGELES BASIN HOSPITALS, 1958[3]

| Week | CO | Case fatality | | Hi > Low? | Week | CO | Case fatality | | Hi > Low? |
		High	Low				High	Low	
1	9.56	21.3	50.0	−	27	5.63	29.2	22.2	+
2	9.20	27.4	0.0	+	28	6.58	26.1	23.1	+
3	9.41	30.6	14.3	+	29	5.54	22.9	8.3	+
4	7.52	24.1	0.0	+	30	5.56	20.9	0.0	+
5	6.84	39.1	28.6	+	31	6.97	41.5	7.1	+
6	8.62	30.0	23.1	+	32	6.95	24.4	22.2	+
7	8.43	24.1	44.4	−	33	7.17	23.8	8.3	+
8	6.58	21.6	20.0	+	34	6.11	15.0	15.4	−
9	6.51	32.0	50.0	−	35	5.78	18.6	27.3	−
10	5.78	26.7	14.3	+	36	7.19	33.3	28.6	+
11	5.81	29.5	10.0	+	37	7.33	23.1	13.3	+
12	6.67	23.4	33.3	−	38	8.49	22.9	16.7	+
13	6.58	15.8	45.5	−	39	7.73	33.9	15.4	+
14	5.87	28.3	35.7	−	40	8.30	24.2	0.0	+
15	7.20	25.4	27.3	−	41	7.88	37.2	28.6	+
16	7.04	14.0	30.0	−	42	8.19	22.9	30.0	−
17	5.81	22.5	14.3	+	43	7.64	29.0	18.2	+
18	5.96	24.5	30.8	−	44	9.50	29.6	17.7	+
19	5.57	21.2	58.3	−	45	8.99	29.6	22.2	+
20	5.83	26.5	28.6	−	46	5.64	26.2	28.6	−
21	6.43	25.5	28.6	−	47	11.58	28.9	8.3	+
22	5.96	13.0	15.4	−	48	8.61	28.6	14.3	+
23	5.36	24.2	20.0	+	49	10.46	31.3	25.0	+
24	5.84	45.2	33.3	+	50	10.39	41.0	27.3	+
25	6.19	32.7	26.3	+	51	14.53	58.6	20.0	+
26	6.21	22.5	9.1	+	52	9.31	28.6	8.3	+

promptly and productively studied with results that tended to support the general proposition (see next section).

The California Health Department promptly requested support from the EPA for a follow-up study in which variables could be better specified. The federal agency, in conjunction with the research arm of the Coordinating Research Council, representing the concerns of the motor vehicle and petroleum industries, decided to conduct such confirmatory studies as it would support in Baltimore, which had very low levels of carbon monoxide! Not surprisingly, such a study failed to confirm an association between increased exposures to carbon monoxide and onset of sudden death or myocardial infarction.[7]

IX. CLINICAL AND EPIDEMIOLOGICAL STUDIES TO WHICH THIS STUDY LED

Hexter and Goldsmith in 1971 published a study of daily mortality in Los Angeles County for the years 1962—65.[8] They found that cyclic variation and maximal temperature were the largest contributors to the observed variation. From the residual variation, after these effects were "removed" by using a mathematical technique involving regression analysis, they were able to show a statistically significant contribution of variation of carbon monoxide, but no effect of variation of photochemical oxidant. To avoid the effects of autocorrelation,

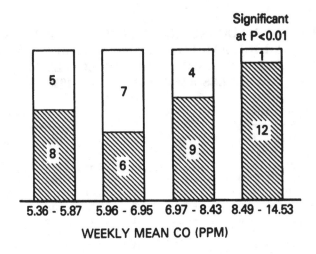

Significant
at P<0.01

WEEKLY MEAN CO (PPM)

Number of weeks with higher case fatality rate

▨ in more polluted area

☐ in less polluted area

FIGURE 3.   Sign test comparing weekly MI admission case fatality
rates in "more polluted area" vs. "less polluted area" by quartile
of weekly mean CO.

they then repeated the analysis for each day of the week and found virtually the same result. Finally, they restricted their analysis to deaths from arteriosclerotic heart disease (which includes deaths due to acute myocardial infarction) and found similar results, which they reported to confirm the findings of case-fatality rate study. While the case-fatality rate study, by its nature could not provide a dose response relationship, the daily mortality study did conclude: "The estimated contribution to mortality for Los Angeles County for an average of 20.2 ppm (the highest concentration observed during the 4-year period) as compared with an average carbon monoxide concentration of 7.3 ppm (the lowest concentration observed) is 11 deaths for that day, all other factors being equal."

Ayres and his colleagues were the initial group of clinical investigators to provide supporting evidence.[9] They showed first that during exercise, the added amount of oxygen needed by the skeletal muscles could be obtained both by increasing the blood flow and by increasing the proportion of oxygen taken out of each unit of blood passing through the muscles; in contrast, the heart muscle (myocardium) in a normal person who is exercising can only increase its oxygen supply by increasing the blood flow in the coronary arteries, since even at rest it extracts from the blood flowing through it about as much oxygen as it can. Thus a person with coronary heart disease (the condition leading to heart attacks) who exercises cannot either increase the proportion of blood oxygen or increase the blood flow, because his coronary blood vessels are either rigid or partly obstructed by plaques on the inside of the vessel walls. Thus, people with coronary artery disease, even without exposures to CO have difficulty in providing the heart muscle with enough oxygen during exercise. They are therefore unusually sensitive to the interference with the oxygen transport function of the blood which CO produces even at low levels of exposure.

Some of the people with coronary artery disease get chest pain when they exercise. This is called angina pectoris, and is a relatively common disease among older persons. Another group of clinical investigators, headed by Dr. W. S. Aronow, had developed methods for

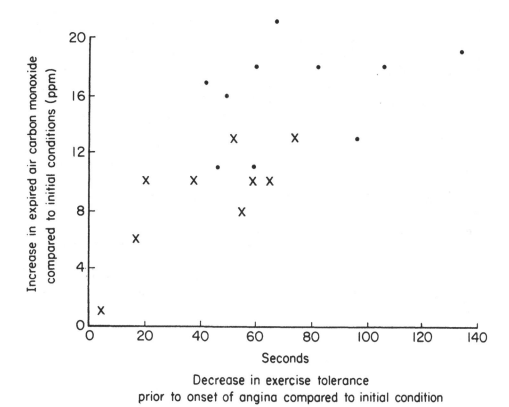

FIGURE 4.   Decrease in exercise tolerance of patients with coronary heart disease manifested by angina pectoris (chest pain on exercise) in relationship to recent freeway exposures the magnitude of which is estimated by the increase in the expired air carbon monoxide compared to original conditions; (●) indicates immediately after exposure and (X) 2 hr later.

estimating just how much exercise patients with angina pectoris could tolerate before getting pain (and the electrocardiographic changes which often go with it). Dr. Aronow and his colleagues working in Los Angeles set out to find out how much effect exposures to carbon monoxide would have on the exercise tolerance ot patients with angina pectoris. In their first experiment, the exposure was simply by driving around Los Angeles on the freeway system.[10] They found that there was a significant decrease in exercise tolerance after such an exposure, a decrease which was not present when the trip was repeated but the subjects breathed purified air with no carbon monoxide. The carbon monoxide exposure was estimated by the change in the expired air carbon monoxide level. Their results are shown in Figure 4. From the figure it can be seen that the effect persists as long as 2 hr after such an exposure. Critics of this study noted that the subjects as well as the investigators could easily tell when the men were exposed, and since there is a subjective element in deciding when chest pain occurs, their knowledge could have biased the results. Accordingly, they did what is called a "double blind" study in which neither the subjects nor the investigators knew whether the men were inhaling carbon monoxide or pure air in the laboratory.[11] These results confirmed that as little as an increase of 1.6% of carboxyhemoglobin can decrease significantly the exercise tolerance of patients with angina pectoris. They also showed in another double blind study that persons with poor arterial circulation to the legs got leg cramps earlier with exercise on exposures to similar levels of CO.[12]

Finally, Anderson, et al. showed the same kind of decreased exercise tolerance for angina patients exposed to CO in a double blind study, and their results were confirmed by the observation of changes in the electrocardiograph as well.[13]

One unexpected result of these studies was the light they shed on the reasons why cigarette smoking leads to such a great increase in deaths and illness from heart disease. Of course cigarette smokers have substantial exposures to CO in the cigarette smoke; the average pack a day smoker has about 5% of the hemoglobin combined with CO, that is has about 5% COHb. Goldsmith and Aronow in reviewing the data concluded that the heart is the most vulnerable organ to low level carbon monoxide exposures, and that such exposures in smokers could explain much of the excess of heart disease which is the largest cause of excess deaths among smokers.[14]

At present (1983), motor vehicle emission regulations in the U.S., Europe, and Japan require devices to substantially reduce CO emissions. Although the extent of control is still contested by the motor vehicle industry, the need for some control is now generally accepted. However, cigarette smoking exposure has not been reduced because scientific concern has been directed to the nicotine and tar content of cigarette smoke and while today's average cigarette gives a reduced exposure to nicotine and tar, there is no certainty that it reduces exposures to CO; in fact for some smokers who smoke more cigarettes with low tar and nicotine content in order to get the same effect, the exposure to CO may even be greater than when they were smoking cigarettes with no filters and higher tar and nicotine.

Indoor heating equipment which is not vented to the outside may be a source of CO exposures as may occupational exposures.

REFERENCES

1. **Goldsmith, J. R.,** Contribution of motor vehicle exhaust, industry, and cigarette smoking to community carbon monoxide exosures, *Ann. N.Y. Acad. Sci.*, 174, 122, 1970.
2. California State Department of Health Technical Report of California Standards for Ambient Air Quality and Motor Vehicle Exhaust, Berkeley, California, 1960.
3. **Cohen, S. I., Deane, M., and Goldsmith, J. R.,** Carbon monoxide and survival from myocardial infarction, *Arch. Environ. Health*, 19, 510, 1969.
4. California Air Resources Board, Ambient Air Quality Standards, Sacramento, California, 1970.
5. U.S. Department of Health, Education and Welfare, Air Quality Criteria for Carbon Monoxide, National Air Pollution Control Administration Publ. No. AP-62, Washington, D.C., 1970.
6. U.S. National Academy of Sciences-National Research Council, Committee on the Biological Effects of Air Pollution, Effects of Chronic Exposure to Low Levels of Carbon Monoxide on Human Health, Behavior and Performance, Washington, D.C., 1969.
7. **Kuller, L. H., Radford, E. P., Swift, D., Perper, J., and Fisher, R.,** The relationship between ambient carbon monoxide levels, postmortem carboxyhemoglobin, sudden death, and myocardial infarction, *Arch. Environ. Health*, 30, 477, 1975.
8. **Hexter, A. C. and Goldsmith, J. R.,** Carbon monoxide: association of community air pollution with mortality, *Science*, 172, 265, 1971.
9. **Ayres, S. M., Muller, H. S., Gregpry, J. J., Gianelli, S., Jr., and Penny, L. J.,** Systemic and myocardial hemodynamic responses to relatively small concentrations of carboxyhemoglobin (COHb), *Arch. Environ. Health*, 18, 699, 1969.
10. **Aronow, W. S., Harris, C. N., Isbell, M. W., Rokaw, S. N., and Imparato, B.,** Effect of freeway travel on angina pectoris, *Ann. Intern. Med.*, 77, 669, 1972.
11. **Aronow, W. S. and Isbell, M. W.,** Effect of carbon monoxide on exercise-induced angina, *Ann. Intern. Med.*, 79, 392, 1973.
12. **Aronow, W. S., Stemmer, E. A., and Isbell, M. W.,** Effects of carbon monoxide exposure on intermittent claudication, *Circulation*, 49, 415, 1974.
13. **Anderson, E. W., Andelman, R. J., Strauch, J. M., Fortuin, N. J., and Knelson, J. H.,** Effect of low level carbon monoxide exposure on onset and duration of angina pectoris, *Ann. Intern. Med.*, 79, 46, 1973.
14. **Goldsmith, J. R. and Aronow, W. S.,** Carbon monoxide and coronary heart disease: a review, *Environ. Res.*, 10, 236, 1975.

Chapter 15

LEAD EXPOSURES OF URBAN CHILDREN: A HANDICAP IN SCHOOL AND LIFE?

John R. Goldsmith and Herbert L. Needleman

TABLE OF CONTENTS

I. THE PROBLEM

Lead poisoning of workers is one of the best studied of all the occupational risks. Colic, paralysis of peripheral nerves, and anemia are the outstanding symptoms; the likelihood of their occurrence is related to the amount of lead stored in the body, an amount which can be estimated by measurement of the lead levels in blood (PbB). In the 1940s it was believed that if the blood level were kept below about 80 μg/100 mℓ in workers, these toxic effects would not occur.

Since early in this century, it had been recognized that small children could be severely intoxicated by eating lead-based paint chips; the symptoms were convulsions, coma, and there was a substantial fatality rate.

These findings led to the widespread restriction of lead-based paint on household furniture and toys, but many older homes built before the risk was well known, still have many layers of high lead paint on surfaces easily reached by toddlers.

One major use for lead was as a motor fuel additive in the form of tetraethyl lead; when the lead-containing gasoline is burned, a fine fume of lead oxide is emitted. Although a fraction of the emitted lead particulate falls to the ground near such streets or highways, some of it is so fine that it drifts about in the air of urban areas. Because of the huge number of motor vehicles used in Los Angeles and the limited capacity for dispersion of emitted pollutants, the Los Angeles Basin was a reasonable location for an initial effort to assess the problem of a possible lead hazard associated with motor vehicle emissions. Guidelines were provided in 1959 when the California State Health Department first proposed scientifically-based air quality standards, and the eminent authority and Medical Director of the Ethyl Corporation, Dr. Robert A. Kehoe gave his opinion that the public's health would be protected so long as the level of lead was low enough so that none would be accumulated in the bodies of the people exposed. At that time some of the levels observed in Los Angeles samples were as high as 12 μg/m^3, but the average of levels was somewhat above 2 μg/m^3. Assuming that the resting adult's air exchange at rest was about 10 m^3 a day, that represented a nontrivial dose.

The California Health Department therefore asked the U.S. Public Health Service to carry out a cooperative study of the possible relationship of blood lead levels and atmospheric lead levels in Los Angeles and several other communities. When this study was carried out, it showed a general tendency for lead in blood of adults to be higher in the locations in which the lead in air was higher; for the most part the high locations were urban and the lower values were found in adjacent rural areas. The mean values forming what was called the three-city study were around 16 μg/100 mℓ of blood for women and around 19 μg/100 mℓ for men. Smokers had slightly higher values than nonsmokers, and a population living in the High Sierra Mountains of California had values of 10 to 12 for men and women respectively.

In 1966, Dr. Kehoe reported on a set of experimental studies of human volunteers who were exposed to increasing amounts of lead by inhalation. Goldsmith and Hexter plotted the results of the Kehoe experiments on the same logarithmic scale as the data from the three-city study, and showed that the two different data sets seemed to cluster about the same line, which they interpreted as showing that with increased air pollution levels above about 2 μg/m^3 there would be increased storage of lead in the body of adult men, Figure 1.[7] Since he felt that the requirement put forth by Dr. Kehoe had been met, Goldsmith proposed that an air qualtiy standard be set to prevent such an accumulation. Figure 1 shows the data presented by them.

In the meanwhile, investigators in Finland and Yugoslavia had shown that low levels of lead exposure in workers could interfere with the body's synthesis of hemoglobin, so that such an effect was suggested as an early sign of biochemically harmful lead exposures.

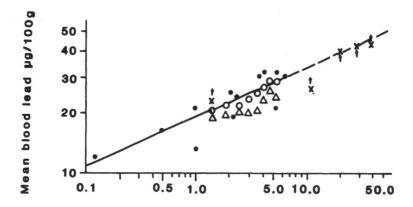

FIGURE 1.   Consistency between dose-response relationships for experimental and epidemiological studies of lead exposures in human subjects. The solid symbols represent the estimated dose and the mean blood lead levels observed in the studies in Cincinnati, Los Angeles, and Philadelphia for different groups. The other symbols represent the achieved blood lead levels of subjects experimentally exposed in the laboratory of Dr. Robert Kehoe. The solid line is the calculated logarithmic regression based on the data from the community studies only. (From Goldsmith, J. R. and Hexter, A. C., *Science*, 158, 132, 1967. With permission.)

Table 1
LEAD CONCENTRATION IN AIR AND IN THE BLOOD OF PRIMARY SCHOOL CHILDREN IN TWO COMPARABLE CALIFORNIA TOWNS (1972)

Community	Lead in air (μg/m³)	Lead in blood of children (μg/100 mℓ)	
		Boys	Girls
Burbank	3.27 ± 1.59(10)	23.3 ± 4.70(17)	20.4 ± 2.91(19)
Manhattan Beach	1.87 ± 1.37(10)	16.8 ± 4.01(21)	17.1 ± 4.37(19)

Note: The data are means and SD with the number of samples in parentheses. The differences between the two communities are statistically significant for both boys and girls, with a probability of $p < 0.01$.

Subsequent work has in general confirmed the nature of the relationship first shown by Goldsmith and Hexter, for many other populations of men, women, and children. Table 1 for example shows data from a study of primary school children from socioeconomically similar census tracts in a heavily polluted area of Los Angeles County (the city of Burbank) and in an area along the coast with relatively little pollution, Manhattan Beach.

The possibility of neurological or psychological impairment of children was raised as a result of examining children in the vicinity of two smelters, one in El Paso, Texas and the other in Kellogg, Idaho. Intelligence tests (I.Q.) and other psychological tests were done in the vicinity of the El Paso smelter by two teams of investigators, one of which was retained by the smelter owner, among somewhat different groups of children at different distances from the smelter, using somewhat different methods. The smelter owner's team found no association with lead exposures, but the other team with University and Public Health input found some evidence that the children with higher lead levels in blood had I.Q. tests which were several points lower. A panel of outside scientists was asked to evaluate the conflicting claims, and it concluded that neither group had used well standardized methods. It recommended that if such studies were done, careful attention must be paid to research design and use of standardized methods. If psychological tests were to be used, it was important that the persons performing the test have no knowledge of the exposure status of child they were testing, so called "blind testing", meaning that the tester was "blind" to whether the child was exposed or not to excess lead levels.

In the vicinity of the Kellogg smelter, evidence of slowing of nerve conduction was obtained for both adults and children.

Another piece of evidence that there may be some connection between lead exposure and mental and psychological development of children was the observation that among children who were hyperactive or had other learning difficulties there were elevated blood lead levels. One difficulty with interpreting these reports was the known inclination of some children to eat nonfood items, an inclination that seems to be stronger in children with developmental problems. So it was possible that the increased blood lead levels were either the result or the cause of the classroom difficulty. That is, children who were neurologically handicapped may also have a tendency to eat nonfood items, a habit called pica, which since many of them contain lead could increase lead burdens; alternatively, lead exposure itself could cause nervous system damage in children just as it causes peripheral nerve damage in adult workers.

From this body of evidence, Dr. Herbert Needleman, a psychiatrist at the Children's Medical Center in Boston put forward the hypothesis that lead levels to which preschool children are exposed may be related to subtle neurological and psychological problems first noted among primary school children. He also proposed and put into effect a unique way of testing the relationship.

II. THE SETTING

The Children's Medical Center is the referral center for Boston and the New England region, as well as a part of the Harvard Medical School complex. It had a long history of study and evaluation of lead toxicity from paint ingestion. A federal program for screening "high risk" children in the region had been under way since 1973. Two communities were chosen from the Boston Standard Metropolitan Statistical Area which had an unusual proportion of the housing units built before 1950, a circumstance which makes it unusually likely that the children had an opportunity to be exposed to lead-based paint. Of the dwelling units in Chelsea and Somerville, 91.2% and 94%, respectively, were built prior to 1950, compared to 68.9% for the whole of Massachusettes.

The 3329 children attending the first and second grades in the period between 1975 and 1978 were the population to be sampled and studied.

III. THE AIMS OF THE STUDY

The study took advantage of what was known about the storage of lead in the tissues of the body. X-rays of children with a history of lead ingestion and poisoning show a band of lead deposition where new bone is being laid down. This pointed to the fact that lead is stored in new bone as it is laid down, and it was shown that blood lead reflects exposure in the recent weeks or months; lead levels in bone reflect exposures in recent years. If one could somehow sample a bony tissue in children, it could be used to indicate the lead exposures during the preceding years.

When school children shed their "baby" (or deciduous) teeth between 6 and 8 years of age, they do in fact offer us an opportunity to sample bone deposition which occurred in preschool years. Between 1972 and 1974, in fact, Dr. Needleman had shown that analysis of the dentine (the inner bony part of the tooth, covered by the harder enamel) of deciduous teeth was a valid reflection of the lead exposures during preschool years.

The study had, therefore two aims:

1. To see if the levels of lead in dentine of a sample of children from a community were associated with any of the psychological tests which were combined to estimate intelligence.
2. To see if there were any association of dentine levels of deciduous teeth with teachers assessment of classroom performance and behavior.

IV. METHODS FOR THE RESEARCH

A. Collection and Classification of Teeth

All of the first and second grade children were asked to submit the teeth they shed to their teacher who then checked to see that there was a fresh socket. The teeth were cleaned, sliced, and the dentine was separated from the enamel in a 1-mm slice. The dentine was then analyzed for its lead content. The children whose initial tooth slice was in the upper 10% of the values for lead (PbT) were initially classified as "high lead" children and the ones with teeth in the lowest 10% were characterized as "low lead" children. Whenever possible, a second sample, either from the same or another tooth was required to confirm that the child was still in the high lead group, and if subsequent values were discordant the child was put into an intermediate group.

B. Neuropsychological Testing

Because the tests depended on English language fluency, children from bilingual homes were not tested, nor were tests used from children whose past history gave any evidence of low birth weight, birth damage, or other reasons for feeling that the child may have a learning disability or have had lead poisoning. About 10% of parents did not want to participate, and about 20% were unable to come to the testing location. Satisfactory tests for children who met all the requirements were scored for 100 children in the low lead group, all of whose PbT values taken together averaged less than 10 ppm, and for 58 "high lead" children all of whose Pbt values averaged over 20 ppm.

The tests began with the Standard Wechsler Intelligence Scale for Children and included seven other well-standardized tests of intelligence, coordination, visual, auditory, and motor competence. The parent, usually the mother, also filled out a medical and social history, an attitude evaluation and had an I.Q. test as well.

All psychological testing was done without the examiners or coders knowing the PbT level; that is, all testing was blind.

C. Teacher's Behavioral and Performance Rating

The teacher of every child who gave a tooth was asked to fill out an 11-item scale concerning classroom performance and behavior. All teachers were blind to lead levels, and were asked to base their evaluation on at least 2 months of classroom experience with the child.

D. Validation Procedures

The teacher's evaluations were compared for the children tested and those excluded, and no systematic differences were found; neither was there a difference of dentine lead level between the excluded group and the group which was tested and analyzed. Thirty-nine other variables which may have affected intelligence and classroom performance were scaled and coded. Parental attitude toward school and child did not differ between the high and low

PbT groups. Only two variables differed between the two groups with respect to the birth and developmental histories of the child, namely pica which was more common in the high PbT group, and the high PbT group was 3.5 months older on the average. Among the parental variables, there were small differences in education, social class, and I.Q. of parents, in the mother's age at the child's birth, and in the number of pregnancies; these were taken into account in the analysis by what is called analysis of covariance.

For some of the children, blood lead levels were available 4 or 5 years prior to the shedding of teeth, so it was possible to validate the premise that high PbT was a reflection of high PbB in earlier years.

V. RESULTS

Of the total population of 3329 eligible children, 2335 submitted at least one tooth, which permitted an analysis of 3221 samples for PbT. Teacher's behavioral ratings were available for 2146 of the children who submitted teeth.

Blood lead levels for 23 of the 58 high PbT children were available; their mean and SD were 35.5 ± 10.1 µg/dℓ 4 to 5 years earlier; for 58 out of the 100 low PbT children, blood lead tests were available and their PbB levels were a mean of 23.8 ± 6.0 µg/100 mℓ.

Children with high PbT levels performed significantly less well on the Wechsler I.Q. Test, particularly on the verbal items. Their performance was also significantly poorer on measures of auditory and verbal processing, and reaction times with delay (Table 2). The teacher's performance evaluation showed a general gradient of deterioration with increasing lead levels in dentine, Figure 2.

In the former set of psychological and I.Q. tests, the potentially confounding co-variates were taken into account, but for the teachers' assessment studies no such adjustment was made. Since pica is three times more frequent in the high PbT group than in the low lead group, a separate analysis must be made, for which the results are shown in Table 3. As can be seen, there is a significant difference for the group without pica, but for the pica group, the numbers are small, and no sigificant differences can be shown based on a $p <$ 0.05, two-sided test.

VI. INTERPRETATION

Because of the use of standardized testing procedures, the blinding of the observers and raters of the test results, and the number of co-variates checked, and taken account of, and the agreement between the teacher's ratings and the psychological test results, this study has provided very powerful evidence of a possible role of lead exposures in early childhood in handicapping school performance and intelligence. The study tends to support the possible role of lead as the cause rather than the result of developmental handicap in that the indicator of lead exposure is one which reflects lead exposures which occurred years ago. Because the effect does not appear to be very great, this probably accounts for the fact that it has apparently been overlooked for so long. On the other hand, the long-term implications of even small deficits in learning capabilities at this age are very great, so we will naturally want to see if there are any other confirmatory studies.

A. Other Studies which may Help to Interpret these Findings

First let us look at the results of a nationwide sample of children in order to see what the patterns of elevated levels of lead may be. Table 4 shows lead levels are highest in black, urban children, and lowest in urban white children. The U.S. Center for Disease Control has designated a level of 30 µg/dℓ of blood (in combination with 50 to 250 µg/dℓ of free erythrocyte protoporphryn, FEP) as evidence of "undue lead absorption". The data in Table 4 were obtained during 1976—1980 by U.S. Department of Health and Human Services'

Table 2
TEST RESULTS FOR FIRST AND SECOND GRADE CHILDREN OF THE POSSIBLE RELATIONSHIP OF LEAD LEVELS IN DECIDUOUS TEETH AND PSYCHOLOGICAL TESTS

Test	Low lead group mean	High lead group mean	Significance[a]
Full scale I.Q.	106.6	102.1	0.03
Verbal I.Q.	103.9	99.3	0.03
Performance I.Q.	108.7	104.9	0.08
Seashore rhythm	21.6	19.4	0.002
Sentence repetition	12.6	11.3	0.04
Reaction time (12 sec delay)	0.41	0.47	0.001

[a] These are two- sided tests, treated as individual results, and co-variables of mother's age at the child's birth, mother's educational level, father's socioeconomic status, number of pregnancies, and parental I.Q. have been taken into account.

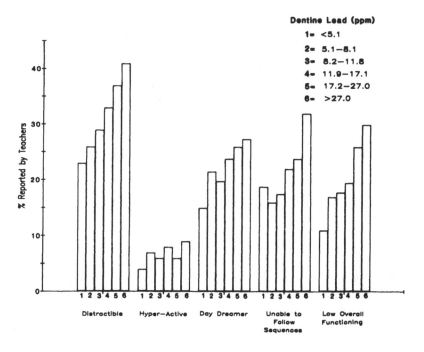

FIGURE 2.   Selected behavioral traits blindly scored by teachers for children in Chelsea and Somerville primary schools according to the dentine lead levels of the deciduous teeth, classed in groups from 1 to 6 as shown in the figure.

National Health and Nutrition Survey. But during the same period, there was a substantial reduction of lead used in motor fuel, and an associated sharp drop in both atmospheric lead levels and also in lead levels in the blood of preschool children, Figure 3.

In Great Britain, a study of 6- to 12-year-old children's blood lead levels in association with I.Q. tests showed significant associations of lower tests with high blood lead, after parent's social class was adjusted for co-variance analysis. Another British study examined lead levels in teeth, psychological tests, teacher's ratings in light of socio-economic class,

Table 3

STRATIFICATION FOR PICA IN ANALYSIS OF VERBAL AND PERFORMANCE I.Q. TESTS FOR FIRST AND SECOND GRADE CHILDREN IN ASSOCIATION WITH LEVELS OF DENTINE LEAD LEVELS

Test	PICA	Low lead	High lead	Significance
Verbal I.Q.	+	101.8 (10) ± 12.0	99.2 ± 13.6 (15)	ns
	−	105.1 (87) ± 12.8	98.8 ± 12.6 (42)	$t = 2.62$ $p = 0.0091$
Performance I.Q.	+	106.9 (10) ± 12.3	103.4 ± 14.4 (15)	ns
	−	110.7 (87) ± 13.5	106.0 ± 11.7 (42)	$t = 2.01$ $p = 0.044$

Note: ns Means not significant. *t* Stands for "Student's *t* test. Means ± SD, numbers in parentheses.

Table 4

PERCENTAGE OF CHILDREN AGES 6 MONTHS TO 5 YEARS WITH BLOOD LEAD LEVELS GREATER THAN 30 µg/dℓ (NHANES II, 1976 to 1980)

	Race	
Annual Family Income	White	Black
Under $6,000	5.9%	18.5%
$15,000 or more	0.7%	2.8%

and data from home interviews. This study found that PbT levels were related to home cleanliness, to mother's smoking, and to the family's social class. Before controlling or adjusting for the social variables there were significant differences for psychological tests, but not quite significant differences for teacher's ratings; after adjustment for the effects of social variables, the difference in tests became "statistically nonsignificant". The authors hedge their interpretations, saying that "it is uncertain whether measurable improvements in the children's intelligence, educational attainment, behavior, etc. would result from a reduction in the body lead burden of children."

In Germany, two community studies showed very much the same thing, a just significant effect which was diminished by taking social and economic variables into account. These two studies were done in the vicinity of nonferrous metal smelters. In the second study, after adjustment for social and economic variables, the high PbT children scored 4.6 points less on the I.Q. tests than did the low PbT children.

Thus, with five large-scale studies in three countries showing very similar results, it is hardly possible to doubt that there is some biological basis for impairment of the psychological competence of school children by lead exposures in early years of life. It is possible that the effect is greater among poor and underprivilged, but if so its consequences are likely to be no less important.

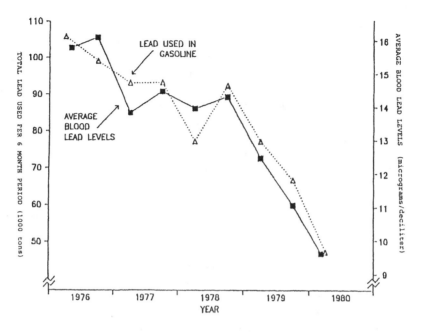

FIGURE 3. Lead used in gasoline production and average NHANES II blood lead levels (February 1976 to February 1980).

We have some evidence as to mechanism of uptake of lead by urban preschool children; in addition to the two obvious ones of inhalation of air polluted with lead from smelters or motor vehicle exhaust and ingestion of lead-based paints, there is recent evidence that when children play in dirt with a high lead content, they get dirt into their mouths from their hands. Since lead from exterior as well as interior paint contributes to lead levels in dust in and around houses, that is one possible source; we have already noted that much of the lead in motor fuel additives is emitted with the exhaust and becomes part of the street dust. It follows that urban children who have no parks or gardens to play in, and as a result play in the street or in dirt near streets or houses painted with lead-based paint would have a greater lead intake than would children whose play is in cleaner environments. In a recent controlled trial of reducing household dust exposures for children 15 and 72 months of age who had elevated blood lead levels, the average blood lead level dropped from a mean of 37.6 μg/dℓ to 31.7 μg/dℓ, with the children having the highest blood lead level at the outset showing the greatest drop. A comparison group which had usual methods of removing lead-based paint used in their homes did not show a drop in blood lead. It is of course much more likely that children from low income urban families will play in street dust, will pick up lead on their hands, and will thereby ingest more lead than will children from better-off families.

VII. THE REGULATORY CONSEQUENCES OF THESE FINDINGS

In the 15 years since air quality regulations were first proposed and adopted by the State of California, the evidence that there are serious health consequences of lead exposures has grown more compelling, and today the evidence that the learning potential of low income, urban children can be handicapped by their playing in a lead-polluted environment should motivate the regulatory arm of governments to take a strong preventive role.

A British Royal Commission has recommended that lead be removed from petrol (gasoline) by 1990 in conjunction with other European countries in the European Economic Community. Although substantial progress has already been made in lowering the lead levels in motor fuel in the U.S., a recent statement of the U.S. Environmental Protection Agency casts doubt on the statistical methods used by Needleman and in particular to question whether if other methods of data selection and of analysis had been used, statistical significance would have been reached. There is no magic associated with the conventional levels of statistical significance, and an effect that appears to have a likelihood of being due to chance of 0.06 is not that much different than one with a probability of 0.04. But the likelihood that five entirely different studies in three different countries would all have effects of about the same magnitude and about the probability is most unlikely to be due to the effects of chance. Such a result could occur if a similar type of bias affected each of the studies, but even the most skeptical observer has not alleged this.

Although lead exposures originating in older housing must have played a larger role in this set of community studies in Chelsea and Somerville than it may play in low income areas of New York or Los Angeles, there is no reason to believe that lead ingestion from motor vehicle exhaust will have any different effect than will lead derived originally from paint, particularly if both of these sources yield dust which contaminates play areas as the common route of exposure.

Restricting the use of lead-based paint, restricting the use of lead additives in motor fuel and provision of clean play areas for urban preschool children are all indicated in the effort to eliminate the handicapping effects of lead exposures on learning.

In both Great Britain and the U.S., scientifically oriented community-action groups have played a major role in urging governmental regulatory action. The Environmental Defense Fund in the U.S. has waged a legal battle to require that the Environmental Protection Agency provide protection from excess lead exposures. In Great Britain, the organization is known as CLEAR, and it has provided expert testimony and letters to scientific journals.

REFERENCES

1. **Needleman, H. L., Gunnoe, C., Leviton, A., Reed, P., Peresie, H., Maher, C., and Barrett, P.,** *New Engl. J. Med.,* 300, 689, 1979.
2. **Ratcliffe, J. M.,** *Br. J. Prev. Soc. Med.,* 31, 258, 1977.
3. **Yule, W., Landsdown, R., Millar, I. G., and Urbanowitz, C. Z.,** *Dev. Med. Child. Neurol.,* 23, 567, 1981.
4. **Smith, M., Delves, H. T., Landsdown, R., Clayton, B., and Graham, P.,** *Dev. Med. Child. Neurol.,* 25, (Suppl. 47), 1983.
5. **Winnecke, G., Hrdina, K. G., and Brockhaus, A.,** *Int. Arch. Occup. Environ. Health,* 51, 169, 1983.
6. **Winnecke, G., Kramer, U., Brockhaus, A., Ewers, U., Kujanek, G., Lechner, H., and Janke, W.,** *Int. Arch. Occup. Environ. Health,* 51, 231, 1983.
7. **Goldsmith, J. R. and Hexter, A. C.,** Respiratory exposure to lead: epidemiological and experimental dose-response relationships, *Science,* 158, 132, 1967.

Chapter 16

LONG TERM HEALTH EFFECTS OF PHOTOCHEMICAL AIR POLLUTANTS; COMMUNITY STUDIES IN THE LOS ANGELES BASIN

S. N. Rokaw and R. L. Detels

TABLE OF CONTENTS

I. BACKGROUND

Photochemical air pollutants differ from other types of air pollutants in etiology, in the kinds of health impact, and in the intensity with which they have been studied. Photochemical (or oxidant) pollution results from atmospheric interactions, energized by sunlight, of hydrocarbon pollutants with nitrogen oxides (see Chapters 9 and 10 for additional background information). In Los Angeles, these derive predominantly from automotive emissions. Nitrogen oxides are also emitted from other combustion sources such as power plants and industrial smokestacks. The mixture produced includes ozone and other highly reactive oxidants which are associated with eye and respiratory tract irritation, characteristic forms of plant damage, distinctive odor, and fine light-scattering submicron-sized particulate matter. The concentrations and duration of substances at ground, or "breathing zone" levels depend on three factors: emissions, atmospheric reactions, and dilution or dispersion in the air mass.

From animal studies and experiments with humans, we have learned that ozone is not solely responsible for all the symptoms and effects associated with photochemical pollution. Aldehydes, olefins, and other complex substances such as peroxyacetyl nitrate compounds and other hydrocarbons are important ingredients of the mixture. These may be respiratory irritants or eye irritants, and they produce characteristic types of damage in controlled experiments with susceptible plants. Ozone is the most readily measured component and therefore is used as a surrogate substance for estimating levels of the total photochemical oxidant mix.

From early experiments, it had been deduced that ozone (photochemical oxidant) exposure would affect human lung function performance with effects that were usually reversible within minutes to hours. Repeated exposures to such irritants were thought to lead to increased cough and sputum, possible permanent decreases of pulmonary function, or aggravation of existing chronic lung conditions such as asthma, bronchitis or emphysema. In 1971, at the time our long term studies were planned, the published data supported the following conclusions:

1. At times and locations with increased photochemical pollution (ozone above 0.15 ppm), eye irritation, respiratory irritation, increased cough and sputum, and possible headache were observed.
2. A fraction of the population with existing asthma was more likely to experience asthma attacks when pollution levels exceeded an hourly average of 0.20 ppm of ozone (see Chapters 9 and 10).
3. Men with chronic bronchitis or emphysema and impaired lung function experienced further lung function impairment when levels of ozone were elevated from 0.10 to 0.30 ppm (hourly average).
4. Mortality from chronic respiratory conditions was showing a general upward trend, but effects of cigarette smoking, survival from tuberculosis and pneumonia, and occupational exposures were considered the predominant contributors. These trends were similar in many urban areas and not appreciably greater in areas with photochemical pollution.
5. Morbidity studies using standard survey methods which had been developed by the British Medical Research Council (BMRC), (see also Chapter 11), pointed to increased symptoms, among both smoking and nonsmoking outdoor workers in Los Angeles, compared to San Francisco, but Los Angeles populations seemed more likely to report symptoms of all kinds than were populations in San Francisco.
6. High school cross-country track teams had "poorer" times when they exercised during

smoggy weather than the same teams exercising when smog was less (range of maximum hourly average values — 0.03 to 0.345 ppm.) (See also Chapter 1.)

7. Experimental animal studies strongly suggested that long term or repetitive exposures to ozone at concentrations relevant to those found in Los Angeles could have significant physiologic or pathologic effects; but only one medium term study had been done in humans. It suggested that after repeated ozone exposures, over several weeks, there was a progressive decrease in the Forced Expiratory Volume at one second ($FEV_{1.0}$) which recovered to the prechallenge values after about 2 weeks.

8. Exposing human subjects to levels of ozone which might lead to irreversible effects could not be ethically done. If persisting effects of photochemical exposure were to be identified, it was essential to undertake community population studies, employing a polluted area as a natural laboratory.

II. CONCEPTS

Methods for community studies relating chronic respiratory conditions to weather and pollution, derived from the BMRC were well known, and criteria for defining initiation and exacerbation of Chronic Obstructive Respiratory Disease (CORD) had been described. The Los Angeles area was most appropriate for the development of long term, community-based studies, since it had been heavily impacted by photochemical oxidant pollution for more than 25 years. Its population was aware of and very concerned with the problem. Many of those who changed their place of residence cited, after employment or economic causation, air pollution as their chief motivation.

A voluntary health agency concerned with respiratory disease (American Lung Association of Los Angeles County) with many decades of community service, primarily directed at control of tuberculosis, had in recent years begun a program of screening for lung function abnormality and respiratory symptoms in both industrial and community settings. Of over 150,000 persons tested, abnormality rates ranging from 10 to 20% had been reported.

A uniform air quality monitoring network, had been in place for more than 20 years providing historical and on-going measurements in the several communities of this large county. Because of topographic, meteorologic, and emission source differences, pollution contrasts existed between areas in the Los Angeles basin permitting selection of populations for comparative studies.

A convergence of these several technologic and public educational skills (Lung Association and Air Quality Management District) with the expertise of the Division of Epidemiology, School of Public Health, UCLA, made it feasible to study, both cross sectionally and prospectively, factors which affect CORD. This chapter deals with the cross-sectional component of the study.

III. OBJECTIVE

The primary objective was to determine if there were measurable effects on lung function associated with residing in an area exposed to high levels of photochemical oxidants. Because we wanted to be able to generalize the results of this study to other communities exposed to photochemical oxidants, it was necessary that the population studied include as many residents as possible and not be restricted to those with respiratory complaints, symptoms, or illness.

A second important consideration was to identify factors in the study populations, other than exposure to oxidants, which might affect lung function, such as cigarette usage, or occupational exposure to substances known to cause respiratory disease. Another concern was that the area with low level pollutants might have attracted a large number of individuals with known respiratory disease, seeking a relatively pollution-free environment in which to live. (In a prior health survey, it had been noted that no adult persons with asthma lived in a most severely polluted Los Angeles location.) Since lung function test performance may differ among various racial groups, it was also important to control for this variable.

Finally, it was necessary to assure that reliable measurement of the levels of air pollution to which the study populations were exposed would be available to merge with the data for individuals.

Although the results reported in this chapter are from the original cross-sectional studies, the ultimate design intention is to follow up the study populations, to determine if rates of change in lung function, or incidence of new cases of disease differed between component groups of the study populations exposed to differing levels of photochemical pollution. Thus, populations were sought which were likely to remain in the study areas, and would continue to be accessible.

IV. METHODS

A. Selection of the Study Populations

We identified two study populations: one, exposed to high levels of photochemical oxidants at place of residence, and a second, exposed to moderately low levels of oxidants, and low levels of other measured pollutants. It was desirable for the study populations to have low mobility as well as similar age distributions, race distributions, and socioeconomic status.

The 1970 census was used to identify census tracts in Los Angeles County which met the criteria listed above. The characteristics of the two selected areas are contained in Table 1. The study populations selected were comparable in the proportion of non-Hispanic whites, median income, and proportion of home owners. In addition, both communities had a high proportion of individuals who had lived in the same household for 5 years previous to the census, suggesting that they constituted a relatively stable population. The decision was made to select entire census tracts, rather than to select a probability sample of a much larger area. This enabled easy access to the test site for participants and closer characterization of the levels of air pollutants to which individuals were exposed. The communities selected were very similar in demographic characteristics, but they had had a differing history of exposure to photochemical oxidants. We were also able to use established air monitoring stations within or adjacent to the study areas to estimate the levels of air pollutants.

The study area exposed to low levels of pollutants was located in Lancaster (see Figure 1). A South Coast Air Quality Management District (SCAQMD) monitoring station was located within that census tract. The study area exposed to high levels of photochemical oxidants was located in Glendora, approximately 4 to 6 km downwind from the Azusa monitoring station of the SCAQMD. The SCAQMD (including the former Los Angeles County Air Pollution Control District) maintained records of measurements for ozone (oxidant), sulfur and nitrogen dioxide, particulates and carbon monoxide, along with the usual meterological measurement of temperature, humidity, and wind direction. Equipment in each of the stations was identical or comparable, and a well-validated calibration and cross checking system had been in place for many years. As new methods for measurement of the several pollutants are introduced, it was cross calibrated with the existing systems. When new systems were substituted, data translations were made using appropriate correction factors. In addition, supervision of this sytem, reporting of data, and computer analysis was reviewed continuously by the California Air Resources Board (CARB) to enable comparability with other communities systems. Thus, it was felt that these monitoring stations would provide a reliable history of oxidant exposure of the residents as well as on-going documentation during the life of the research study. California Air Resources Board investigations confirmed, using both direct measurement and computer modeling that the fixed

Table 1
DEMOGRAPHIC CHARACTERISTICS OF STUDY CENSUS TRACTS IN LANCASTER (LOW POLLUTION) AND GLENDORA (HIGH POLLUTION)[a]

Characteristics	Lancaster	Glendora
Total residents, all ages	7,069	4,573
White (non-Spanish surnamed)	6,430 (90.9%)	4,281 (93.6%)
Spanish-surnamed	434 (6.1%)	162 (3.5%)
Black	91 (1.3%)	3 (0.1%)
Other	114 (1.6%)	127 (2.8%)
Total 7+ years of age	6,121	4,061
Median income	$11,631	$12,746
Median home value	$18,600	$23,850

[a] According to the 1970 census.

stations appropriately estimated levels of photochemical oxidant, sulfur dioxide, oxides of nitrogen, and particulate matter for the study areas.

B. Recruitment of Participants

Having selected the study sites, the next step was to enlist the cooperation of as many of the residents of the selected study areas as possible. A variety of strategies was used to maximize the participation of residents of the study area.

In order to provide credibility to the study, an approach was first made to civic leaders in each community. Next, the local newspapers were asked to print stories about the proposed study which explained the objectives of the study, how participants would be approached, and the nature of the tests which would be administered.

The recruitment strategy adopted was to obtain a listing, by name of the heads of household of all residences in the study area. These lists were constructed from voter registration files, reverse directories, and commercial mailing firms.

A letter was sent to the head of the household which explained the study as a "smog effects evaluation program" and indicated that participants would be contacted by a neighborhood representative. Also enclosed was a copy of the article which had appeared in the local newspaper. The letters were sent out to participants over the period of 10 to 12 months in a sequence which would assure that participants were contacted by the neighborhood representative within one week of receiving the letter. Neighborhood representatives were recruited from among residents of the general community in which the study population was located in order to increase their acceptability to potential participants and because of their probable familiarity with the general area. The neighborhood representative was responsible for contacting the household and for constructing a roster of all individuals 7 years of age or older who resided in the household at that time. Appointments were then set up for the participants to complete lung function testing at the Mobile Lung Research Laboratory (MLRL) which was located within walking distance of the resident's home.

V. EQUIPMENT USED AND ANALYTIC STRATEGIES

The MLRL included a motor home and a 40-ft trailer conveniently located in a public area (usually a local school parking lot) in order to make testing as easy and convenient as possible and to increase the probability of their completing studies. The motor home enabled confidential administration of an interview schedule derived from the standardized (NHLBI) questionnaire, closely patterned on the BMRC questionnaire which also included questions on smoking history, residence history, exposure to various chemicals known to be associated

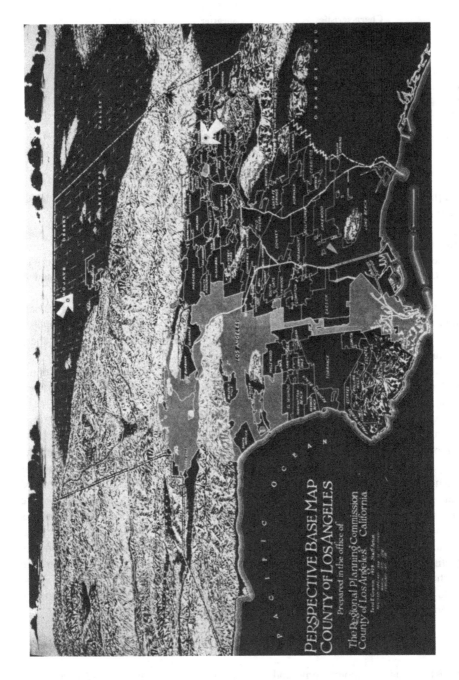

FIGURE 1.   Locations of census tracts in studies of chronic obstructive respiratory disease (CORD). Arrow ○ — census tract in Lancaster; arrow ● — census tract in Glendora.

with respiratory disease, general occupational history, and history of changes of residence or occupation because of respiratory problems. Following completion of the interview schedule, subjects entered the large trailer unit for physiologic testing.

We were concerned that performing only simple spirometric measurement of vital capacity (FVC)* and flow in large airways ($FEV_{1.0}$) might not identify other significant effects on human lung function, such as disturbances in small airways, long term diminution of lung growth and development, or restrictive changes. The tests selected reflected such concerns but respected the need to optimize subject participation. Thus, nontraumatic, noninvasive, quickly performed measures of lung function were employed. Forced expiratory spirometry was augmented by flow-volume analysis of the electronically recorded data. Determination of lung volumes (other than those measured by simple spirometry) and airways resistance were done in a body plethysmograph (volume type). Abnormalities of inspired air distribution and "closing volumes" were investigated, using a single breath nitrogen washout maneuver. Each subject had height, weight, pulse, and blood pressure recorded on entering the unit. After the first year of the program, end-tidal carbon monoxide determinations were secured from each participant, to determine the body burden of CO and to identify recent smoking, which might affect performance of other tests.[1,2]

The field equipment had been cross-calibrated with similar equipment in the Clinical Pulmonary Function Laboratory at UCLA, before field studies were undertaken. Multiple volume and flow calibrations were performed on each day of field testing. Daily validation of the data tapes by the Data Management unit at UCLA School of Public Health minimized losses of data or equipment malfunction or drift. When results had been assembled for each subject, a letter reporting the normality, or extent of divergence from normal was sent to each participant and to his designated physician.

We decided to include only those reports of cough and sputum production which had been present for at least 50 days/year for at least 2 years; or first thing in the morning for as much as 3 months/year for 2 years. Thus, individuals with respiratory symptoms for less than 2 years were not included among individuals "reporting symptoms". Participants were considered to have "chest illness" if they reported having a great deal of trouble with illnesses such as chest colds, tuberculosis or pneumonia, or were unable to perform their usual activities because of chest illness, more than 5 times during the last 3 years. Using these criteria, the prevalence of each of these symptoms was determined in the two-study populations separately for children and young adults 7 to 24 years of age, for adults 25 to 59 years of age, and for individuals older than 59. The populations were divided into individuals who had smoked more than one pack, or less than one pack of cigarettes per day.

Analysis of the results of lung function tests was more complex. In order to estimate the normal forced expiratory volume at 1-sec ($FEV_{1.0}$) forced vital capacity (FVC) and mid-expiratory flow rate between 25 and 75% of observed FVC (FEF_{25-75}), regression equations were developed. These were derived from a population of individuals with no history of smoking and no history of respiratory disease. Expected values were computed based on the age, sex, and height of each subject. The results of lung function testing were analyzed in two different ways. First, the mean percent of expected for the entire group tested was determined. Second, the prevalence of individuals with test results less than 50% of expected for $FEV_{1.0}$, or FVC, or less than 35% of expected for FEF_{25-75} was determined. The mean percent of expected provided an estimate of the general impact of exposure to inhaled agents, while the prevalence figure provided an estimate of the number of individuals whose res-

---

* Appended is a description of lung function tests to help in understanding terms used in this chapter.

piratory function may have been seriously compromised. The results were determined separately for men and women, for never smokers and current smokers, and for specific age groups.

In this chapter, results are reported only for the individuals who were 25 to 59 years of age at the time of testing. Because children grow at different rates it was difficult to evaluate the results of cross-sectional testing in this age group. For individuals in the older age group, there was the added problem that the population over 60 represented a group of individuals who had survived for at least that period of time and would be more likely to suffer from a variety of chronic health problems mostly unrelated to pollutant exposures.

VI. RESULTS

The participants of the Lancaster study area were tested in 1973-74, whereas the individuals in the Glendora study area were tested in 1977-78. Concurrent testing in the two areas was not possible because of funding problems as well as logistics; nor were they studied in consecutive years since this was part of a larger study of four differing study areas. The same test equipment and technical team was used in both areas.

The levels of air pollution extant in the two study areas over the time in which testing occurred are shown in Table 2. Participants in the Lancaster study area were exposed to moderately low concentrations of photochemical oxidants and low concentrations of other pollutants. Participants in the Glendora area (Azusa station) were exposed to very high concentrations of photochemical oxidants and moderately high levels of particulates, nitrogen dioxide and sulfates.

Residents were enumerated in 2143 (84%) of the 2551 occupied households in the Lancaster study area and in 2596 (99%) of the 2629 occupied households identified in the Glendora study area. In Lancaster, 3309 of the occupants of the enumerated households who were 25 to 59 years of age and were non-Hispanic whites participated in testing. Of these, 96% completed both the questionnaire and the lung function tests, 4% completing the questionnaire only. In Glendora, 2647 occupants of the enumerated households participated in the testing, 89% completing all tests, and 11% completing the questionnaire only.

The prevalence of residents with a history of asthma, bronchitis, or emphysema was 11% in Lancaster and 10% in Glendora. Since Glendora was well known to be exposed to high levels of air pollutants, we were surprised to find almost equal prevalences of these diseases. We would have expected individuals who knew they had prior respiratory disease to have avoided moving into, or to have left the Glendora community, and to have been attracted to a community like Lancaster. This assumption may have been confounded by other motivating factors more compelling than health. On the other hand, we cannot eliminate the possibility that at least some of the disease reporting may have been stimulated by the environmental factors present in Glendora. In the analyses which follow, if individuals reported that they had changed residence or occupation because of respiratory problems, they were not included; it was felt that such previously manifested illness might not represent a response to the environment in which they were currently living and being studied.

The mean duration of residence in the study areas did not differ significantly between the areas, nor within age categories between areas. The prevalence of residents who gave a history of working in an occupation which "might be hazardous to respiratory health" was 45% in Lancaster and 32% in the Glendora study area.

The age-adjusted prevalence of individuals who reported symptoms of CORD, stratified by sex and smoking history are shown in Table 3. The prevalence of cough, sputum, and wheezing was consistently higher in Glendora (the area exposed to high levels of photochemical oxidants) both among smokers and never smokers. The frequency of chest illness was greater among men in Glendora, but for women was not consistently greater in one community or the other.

Table 2
ANNUAL MEAN CONCENTRATIONS OF SELECTED AIR POLLUTANTS MONITORED AT THE LANCASTER (LOW POLLUTION) AND AZUSA (HIGH POLLUTION) STATIONS 1972 TO 1977

Pollutant/Area		Mean 72—77	1972	1973	1974[a]	1975	1976	1977[b]
Oxidant[c]	Low	6.5	6.0	8.0	5.7	5.9	6.6	6.5
(pphm)	Hi	11.6	12.3	11.3	12.3	10.8	11.5	11.2
SO$_2$[c]	Low	NA	NA	NA	NA	NA	1.1	1.0
(pphm)	Hi	2.6	3.4	3.0	2.5	2.5	2.1	2.3
NO$_2$[c]	Low	3.2	3.9	3.0	2.8	3.2	3.0	3.5
(pphm)	Hi	11.4	11.8	11.1	11.8	12.0	9.5	12.0
Hydrocarbon[c]	Low	2.9	3.2	3.0	3.0	2.8	2.9	2.8
(ppm)	Hi	4.8	5.6	4.6	4.8	5.0	4.5	4.6
Hi-vol part.[d]	Low	NA	NA	NA	NA	NA	NA	76.0
(mg/m^3)	Hi	NA	NA	NA	NA	NA	NA	133.0
SO$_4$ PARTIC.[d]	Low	NA	NA	NA	NA	NA	NA	4.5
(mg/m^3)	HI	NA	NA	NA	NA	NA	NA	13.5

Note: NA data are not available.

[a] Year of testing in Lancaster, the low pollution area.
[b] Year of testing at Azusa station, the high pollution area of Glendora.
[c] Mean of the daily maximum hourly concentrations.
[d] Geometric mean of the 24-hr total measurements.

Table 4 shows the results of lung function testing. The mean percent of expected and the percent of residents with test results less than 50% of expected for FEV$_{1.0}$, FVC, and less than 35% of expected for FEF$_{25-75\%}$ is shown, again stratified by sex and smoking category. As expected, for these tests, both the mean % of expected and the prevalence of individuals with low test results was worse among smokers than among never smokers. The mean % of expected, for FEV$_{1.0}$, FVC, and FEF$_{25-75\%}$ was greater among Lancaster men than among Glendora men. There was no consistent relationship among the women. On the other hand, the prevalence of individuals with test results less than 50% of expected for FEV$_{1.0}$ and FVC and less than 35% of expected for FEF$_{25-75\%}$ was consistently greater among both male and female residents of Glendora than among residents of Lancaster.

In Table 5, the levels of significance for differences in symptoms and prevalence of low test performance are shown, separately among never smokers, smokers of greater than one pack per day, and smokers of less than one pack per day. (A dash indicates that differences in that test between participants in the two areas was not statistically significant.) In every instance where there is a significant difference between the two communities, the test result was "worse" among Glendora participants than among Lancaster participants. The level of significance was determined using two sample t tests.

VII. ANALYSIS

A number of criticisms could be made of the study described in this chapter.

Some bias might have been introduced as a result of the frank statement of study objectives in introductory publicity and the fact that lung function testing in the two areas was not

Table 3

AGE-ADJUSTED[a] PREVALENCE[b] OF REPORTED SYMPTOMS OF CHRONIC OBSTRUCTIVE RESPIRATORY DISEASE AMONG PARTICIPANTS[c] 25 TO 59 YEARS OF AGE IN LANCASTER AND GLENDORA STUDY AREAS

| | Never smokers | | Current smokers | | | |
| | | | <1 pack | | >1 pack | |
Symptom/sex	Lancaster (M = 574) (F = 736)	Glendora (M = 489) (F = 615)	Lancaster (M = 352) (F = 359)	Glendora (M = 237) (F = 278)	Lancaster (M = 705) (F = 583)	Glendora (M = 481) (F = 367)
Cough						
Men	5.1 ± 1.1[d]	9.7 ± 1.1	22.4 ± 1.9	25.1 ± 1.7	25.9 ± 2.9	28.9 ± 1.8
Women	6.6 ± 1.7	10.4 ± 0.2	24.1 ± 1.6	27.7 ± 2.0	26.5 ± 2.0	30.1 ± 1.9
Sputum production						
Men	4.0 ± 2.0	7.1 ± 1.1	21.1 ± 1.8	23.1 ± 1.4	24.3 ± 2.1	23.8 ± 1.5
Women	7.5 ± 1.9	11.2 ± 1.6	22.9 ± 1.9	24.1 ± 1.0	23.8 ± 1.7	27.2 ± 1.2
Wheezing						
Men	7.2 ± 1.5	13.9 ± 1.6	12.8 ± 1.8	24.1 ± 1.8	13.9 ± 2.2	29.8 ± 1.6
Women	8.2 ± 1.6	14.6 ± 1.1	16.2 ± 2.0	21.5 ± 1.7	17.9 ± 1.9	23.9 ± 2.0
Frequent chest illness						
Men	1.1 ± 1.6	1.9 ± 0.8	2.9 ± 1.7	3.2 ± 1.0	3.3 ± 1.6	4.8 ± 0.9
Women	2.8 ± 0.9	2.6 ± 0.9	9.1 ± 1.5	5.9 ± 1.0	9.9 ± 1.5	7.6 ± 1.3

[a]   Age-adjusted according to 1970 census of the white male or female population of the U.S.
[b]   Percent of all subjects in designated group responding to questionnaire.
[c]   White, non-Spanish-surnamed residents with no history of change of occupation or residence because of respiratory problems.
[d]   Standard error.

concurrent. However, standardization procedures and daily calibration should have reduced the probability that that occurred. In addition, comparison of field and UCLA test results in the same subjects would have identified such errors.

Smoking habits might have changed over the time interval between testing in the two areas. This would have changed the proportion of heavy, light, and never smokers within the population, but we stratified on smoking category in analyzing the results. The changes in the proportion of individuals in the three categories should not have affected the analysis within a category.

It is also possible that other factors were occurring in the communities at the time of testing which could have altered the lung function results of participants. The most likely factor would have been the occurrence of an epidemic of disease (such as influenza) which might affect lung function performance. No such epidemic was observed during the period of testing in either study area.

The study design did not permit an accurate assessment of differences in the proportion of time spent out-of-doors or of differences which might have occurred in the levels of pollutants occurring within the households which might affect lung function. Differences in the proportion of time spent out-of-doors would be expected to reduce the observed differences below those which might have been expected if indeed there is a relationship between levels of photochemical pollutants at place of residence and lung function. Thus, this was a conservative bias and would tend to suggest that the observed differences were an underestimate of real differences between the two communities.

Another criticism which could be leveled at the study is that there were differences in the proportion of individuals who commuted to areas with different levels of air pollutants, in

Table 4

RESULTS[a] OF FORCED EXPIRATORY VOLUME IN 1 SEC, FORCED VITAL CAPACITY, AND MIDEXPIRATORY FLOW RATE IN PARTICIPANTS[b] 25 TO 59 YEARS OF AGE IN LANCASTER AND GLENDORA STUDY AREAS

Test/sex/analysis	Never smokers		Current smokers			
			<1 pack		>1 pack	
	Lancaster (M = 548) (F = 725)	Glendora (M = 477) (F = 597)	Lancaster (M = 306) (F = 348)	Glendora (M = 229) (F = 260)	Lancaster (M = 692) (F = 573)	Glendora (M = 480) (F = 348)
FEV_1						
Men						
Mean per cent[c] expected	105 ± 0.7[d]	103 ± 0.7	104 ± 0.9	102 ± 1.0	101 ± 0.6	98 ± 0.8
Per cent < 50% of expected	0.7 ± 0.3	2.0 ± 0.6	0.7 ± 0.1	1.6 ± 0.6	0.8 ± 0.2	3.0 ± 0.4
Women						
Mean per cent expected	101 ± 0.6	101 ± 0.6	100 ± 1.0	102 ± 1.2	95 ± 0.7	96 ± 0.8
Per cent < 50% of expected	0.6 ± 0.3	2.7 ± 0.5	0.8 ± 0.3	2.5 ± 0.5	1.2 ± 0.2	3.3 ± 0.6
FVC						
Men						
Mean per cent expected	105 ± 0.6	103 ± 0.6	104 ± 0.8	103 ± 0.9	102 ± 0.5	100 ± 0.7
Per cent <50% of expected	0.2 ± 0.1	1.3 ± 0.2	0.5 ± 0.1	1.4 ± 0.4	0.6 ± 0.1	1.8 ± 0.6
Women						
Mean per cent expected	102 ± 0.5	103 ± 0.6	101 ± 0.9	104 ± 1.2	99 ± 0.7	100 ± 0.7
Per cent < 50% of expected	0.7 ± 0.1	1.8 ± 0.2	0.8 ± 0.1	2.3 ± 0.5	1.0 ± 0.2	2.9 ± 0.5
$FEF_{25-75\%}$						
Men						
Mean per cent expected	102 ± 1.3	100 ± 1.3	102 ± 1.8	93 ± 1.9	96 ± 1.1	92 ± 1.5
Per cent < 35% of expected	0.1 ± 0.4	0.4 ± 0.3	1.2 ± 0.5	1.4 ± 0.7	2.2 ± 0.5	2.8 ± 0.7
Women						
Mean per cent expected	100 ± 1.0	97 ± 1.1	97 ± 1.8	96 ± 2.1	87 ± 1.2	89 ± 1.3
Per cent < 35% of expected	0.1 ± 0.4	0.8 ± 0.4	1.2 ± 0.6	1.8 ± 1.0	4.6 ± 0.9	4.5 ± 1.0

a   Age-adjusted according to 1970 census of the white male or female population of the U.S.

b   White, non-Spanish-surnamed residents with no history of change of occupation or residence because of respiratory problems.

c   Of all subjects in designated group completing test.

d   Standard error.

Table 5
DEGREES OF SIGNIFICANCE[a] FOR DIFFERENCES BETWEEN LANCASTER AND GLENDORA STUDY AREAS (SYMPTOM PREVALENCES AND LUNG FUNCTION TEST RESULTS OF PARTICIPANTS 25 TO 59 YEARS OF AGE)

Symptom[b]/test	Never smokers	Current smokers	
		< 1 pack	> 1 pack
Cough			
Men	0.001	—[c]	—
Women	0.005	0.05	—
Sputum production			
Men	0.05	—	—
Women	0.05	—	0.05
Wheezing			
Men	0.001	0.001	0.001
Women	0.001	0.005	0.005
Frequent chest illness			
Men	—	—	—
Women	—	0.05	—
Spirometry			
FEV_1[b] < 50% of expected			
Men	0.01	—	0.001
Women	0.001	0.001	0.001
FVC_1[b] < 50% of expected			
Men	0.001	0.005	0.01
Women	0.001	0.001	0.001
$FEF_{25-75\%}$[b] > 35% of expected			
Men	—	—	—
Women	—	—	—
V_{max}[d]			
Men	0.001	0.001	0.001
Women	0.001	0.001	0.001

[a] Two sample *t* tests were calculated. Probability values in these tests should be interpreted cautiously because sample sizes and rates may be too small to assume normality of the estimate.
[b] Prevalences.
[c] $p > 0.05$ not reported.
[d] Between mean values.

which they spent the majority of their waking hours. In order for this bias to produce the difference observed, it would have been necessary for residents of Lancaster to commute to an area of lower air pollution and for residents of Glendora to commute to an area of greater air pollution.

In fact, only 1% of the Lancaster residents commuted and most of those went to areas with higher levels of air pollution, whereas 23% of Glendora residents commuted, most to areas of lower air pollution (since Glendora is among the communities experiencing the highest levels of pollutants in southern California).

These facts would suggest that the observed differences may really have been an underestimate of the real differences rather than an overestimate.

The differences between the two study areas were greater among men than among women, yet a higher proportion of women are likely to remain in the area of their residence than are men. Other than the possibility that women react less strongly to air pollutants, we were not able to identify a factor which would have increased the probability of observing such a result. However, it has been generally assumed that women spend a larger proportion of time indoors than do men. Screens, walls, carpets, and other common indoor surfaces tend to absorb and neutralize oxidants.

It was conceivable that a number of individuals had migrated to Lancaster because they had respiratory problems and were seeking an area exposed to lower levels of pollutants. From the questionnaire, we identified individuals who had reported that they had migrated to Lancaster and Glendora because of respiratory problems. These individuals were eliminated from the comparisons above. On the other hand, Glendora was well known to be in an area chronically exposed to high levels of pollutants. It is quite probable that susceptible individuals would not have migrated to Glendora because of the high levels of pollutants, and further, that some individuals who developed respiratory disease had moved away from Glendora. Unfortunately, it was not possible to determine the proportion of individuals who had left Glendora because of respiratory problems before the study began. Again, the direction of this bias would have been to reduce the observed differences and therefore to have caused the observed differences to underestimate rather than overestimate the real differences.

A further problem is to separate out acute effects associated with lung function testing on days when high levels of air pollution are occurring from those effects which are the result of long term exposure to photochemical oxidants. Lung function testing in both areas was carried out over a course of 10 to 12 months and individuals were tested during both clean and "smoggy" days. It is not possible to eliminate the possibility that the differences observed may have been due, at least in part, to reduced performance occurring acutely as a result of high air pollution levels at the time of testing. However, differences between the communities were observed among residents tested during the season of lower levels of air pollution as well as during the season when high levels of air pollution occurred and the magnitude of the difference was similar in all seasons.

Cross-sectional studies such as those presented in this chapter are best viewed as a first step in identifying a relationship between exposure at place of residence to high levels of photochemical oxidants, and reduced levels of lung function. These studies should be confirmed by cohort follow-up studies to determine whether the rates of change in lung function over time are greater among residents exposed to high levels of photochemical oxidants at their place of residence than among residents of a clean area.

VIII. IMPLICATIONS

The pathways to CORD are many. Cigarette smoking, history of repeated respiratory infections in earlier years, history of repeated asthmatic exacerbations, exposure to substances encountered in the workplace, genetic susceptibilities, and residence in polluted urban environments have been implicated in the development of this major health problem.

These studies suggest that residing in an area of high photochemical oxidant pollution is capable of contributing to such an outcome. A possible risk of developing CORD is attributable to residing in Glendora, as against Lancaster, indicated by the significant difference in respiratory symptoms and the prevalence of impaired $FEV_{1.0}$ and FVC in adult never smokers. There is not a clearly significant difference in the reported frequency of diagnoses of established obstructive respiratory diseases (CORD), i.e., advanced stages in the course of these diseases have not yet been achieved. Our findings relate largely to functional abnormalities, still short of diagnosed disease. It is not yet clear whether this risk is universal or may be limited to an as yet undefined susceptible subgroup of the population. Those who perceive a health risk to themselves or their offspring regard the development of such data as vital for their decision-making regarding continued residence in such areas or in selecting "safe" areas in which to reside when first they migrate to California. Reports from this

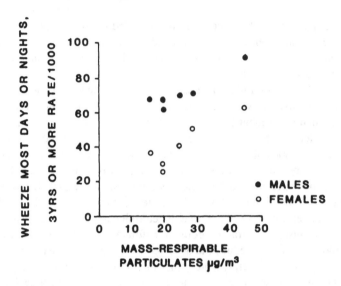

FIGURE 2. In general population samples selected from six communities
with varying levels of pollution, the strongest association was between the
average concentration of mass respirable particulate matter and the symptom of wheezing most days or nights for 3 years or more. As the figure
shows, there appeared to be an interesting association for both male and
female respondents, suggesting that the relationship was not solely due to
occupation, in which case, the association should be strongest for males.[6]

study have not yet had an opportunity to impact such decision making. There are some who
are concerned that respiratory symptoms and reduction of lung function do not inevitably
predict the development of chronic respiratory disease (CORD) and they fear that emphasizing
them will have more unfavorable than beneficial effects. For example, they are concerned
lest there be unwarranted detriment to property values and discouraging of the population
flow needed to do the work of the megalopolis. Underpinning all this is a legal mandate in
the State of California that air quality standards shall be established for the protection of
human health and well being. Data derived from long term, prospective studies provide a
major dimension to the criteria on which such air quality standards are promulgated.

A large population base (greater than 15,000 persons) was initially recruited for these
studies; meticulous care was employed in collecting data, calibrating the measuring equipment, and collating and analyzing the data. The data base is fundamental for the next phase
of the work: prospective studies in which the progression of unfavorable effects of long-
term oxidant exposures, and the incidence of respiratory disease in these populations can
be evaluated.

IX. OTHER RELATED STUDIES

Within this data base is information useful for a number of additional studies, some of
which are already under way. These include studies to estimate the consequences of growing
up in a household in which parents smoke cigarettes as compared to households in which
no one smokes; the patterns of relationship between development of lung function and other
components of human growth; and the attempts to identify subgroups who are unusually
susceptible to inhaled irritants.

Similar related studies were used in a set of community studies primarily focused on
possible long term effects of sulfur dioxide and particulate pollutants in six communities,
Portage, Wisconsin; Topeka, Kansas; Kingston-Harriman, Tennessee; Watertown, Massa-

chusetts; St. Louis, Missouri; and Steubenville, Ohio (in approximate order of increasing severity of the pollution exposures). The presence of wheezing among both males and females in these six communities appeared to be associated with the levels of "Mass Respirable Particulate Matter". By this is meant the relative amount of suspended particulate matter which is of a small enough particle size (usually smaller than 5 μm) to be likely to be lodged in the lung tissues.

A study of nonsmoking members of the Seventh Day Adventist Church in the Los Angeles Basin relating the place of residence to adjacent monitoring stations has also shown an association between presumed exposure to pollution and chronic respiratory symptoms. The association is strongest for Total Suspended Particulate Matter and for Oxidant in that order.

Evidence to date does not indicate any excess cancer risk associated with oxidant exposures,[4] but studies to correlate cancer with air pollutant exposures are particularly difficult to carry out since the relevant exposures may have occurred decades before the diagnosis of the malignant disease. Further, the incidence of lung cancer or other types of cancer is low enough to require follow-up of very large populations.

APPENDIX

PULMONARY FUNCTION TERMINOLOGY

Static Lung Volumes

Tidal volume (VT): Volume of air inspired and expired at each breath.

Inspiratory reserve volume (IRV): The maximum volume of air that can be inspired after a normal inspiration.

Total lung capacity (TLC): The volume of air that can be contained in the lungs at maximum inspiration.

Expiratory reserve volume (ERV): The maximum volume of air that can be additionally expired after a normal expiration.

Residual volume (RV): The volume of air remaining in the lungs after a maximum expiration.

Functional residual capacity (FRC): The volume of air in the lungs at the resting expiratory level (FRC = RV + ERV). Determined by helium dilution technique or by body plethysmography.

Forced vital capacity (FVC): The maximum volume of air that can be forcibly expired after maximum inspiration (TLC = FVC + RV).

Dynamic Lung Volumes and Flow Rates

Forced expiratory volume at 1 sec ($FEV_{1.0}$): Volume expired during the first second of FVC — it normally comprises >75% of FVC.

Forced expiratory volume at 3 sec ($FEV_{3.0}$): Volume expired during the first 3 sec of FVC — usually >92% of FVC.

Forced expiratory flow, midrange ($FEF_{25-75\%}$): Mean expiratory flow over the middle half of the FVC. May be a more sensitive test than $FEV_{1.0}$ for detecting early airways obstruction.

Diffusing capacity (DL_{co}): The number of milliliters of carbon monoxide absorbed per minute per mmHg. Determined by patient inspiring maximally a gas containing a known small concentration of carbon monoxide, hold breath for 10 sec, then slowly exhale to residual volume. An aliquot of alveolar (end expired) gas is analyzed for carbon monoxide to determine the amount absorbed during the breath.

REFERENCES

1. **Detels, R., Rokaw, S. N., Coulson, A. H., Tashkin, D. P., Sayre, J. W., and Massey, F. M., Jr.,** The UCLA population studies of chronic obstructive respiratory disease. I. Methodology and comparison of lung function in areas of high and low pollution, *Am. J. Epidemiol.*, 109, 39, 1979.
2. **Tashkin, D. P., Detels, R., Coulson, A. H., Rokaw, S. N., and Sayre, J. W.,** The UCLA population studies of chronic obstructive respiratory disease. II. Determination of reliability and estimation of sensitivity and specificity, *Environ. Res.*, 20, 403, 1979.
3. **Rokaw, S., Detels, R., Coulson, A., Sayre, J., Tashkin, D., Allwright, S., and Massey, F.,** The UCLA population studies of chronic obstructive respiratory disease. III. Comparison of pulmonary function in three communities exposed to photochemical oxidants, multiple primary pollutants, or minimal pollutants, *Chest*, 78, 252, 1980.
4. **Detels, R., Sayre, J., Coulson, A., Rokaw, S., Massey, F., Tashkin, D., and Wu, M.,** The UCLA population studies of chronic obstructive respiratory disease. IV. Respiratory effect of long-term exposure to photochemical oxidants, nitrogen dioxide, and sulfates on current and never smokers, *Am. Rev. Resp. Dis.*, 124, 673, 1981.
5. **Ferris, B. G., Jr., Speizer, F. E., Bishop, Y. M. M., Spengler, J. D., and Ware, J. H.,** The six-city study. A progress report, in *Atmospheric Sulfur Deposition: Environmental Impact and Health Effects*, Shriner, D. S., Richmond, C. R., and Lindberg, S. E., Eds., Ann Arbor Science-Butterworth, Woburn, Mass., 1983, chap. 13.
6. **Goldsmith, J. R.,** Health effects of air pollution, in *Air Pollution*, Vol. 6, Stern, A. C., Ed., Academic Press, New York, in press, chap. 6.

Chapter 17

MOTOR VEHICLE ACCIDENTS: MICHIGAN CUTS TOLL FROM YOUTHFUL DRUNK DRIVERS

John R. Goldsmith, Alexander C. Wagenaar, and Allen B. Rice

TABLE OF CONTENTS

I. THE PROBLEM

In the U.S., mortality of "young adults" (those aged 15 to 24) is among the highest in the world (179.3/100,000/year for young men and 62.5/100,000 for young women in 1975 to 1978). Some other countries with high per capita income have death rates nearly as high, but other high income countries have relatively low fatality rates (about 100/100,000 for young men). Further data are given in the Appendix.

Deaths due to "external" causes, such as unintended injuries ("accidents"), poisoning, suicide, and homicide are what make the big difference. Among these external causes, the role of motor vehicle crash fatalities is first in frequency and is prominent in the difference between high-fatality-rate and low-fatality-rate countries.

Such deaths not only destroy young people in whom we have invested years of love, nourishment, education, and hope, but they also deprive our communities of many years of potentially productive life. In this sense, these lost lives are a serious economic loss as well. What can communities do that will be effective in cutting down on this tragic waste?

Age alone is a major risk factor for driving, with, for example 16-year-old male drivers having over 10 times the rate of fatal crashes (on a per mile driven basis) than men of their father's ages would have; the relative rates are almost as steep for women, as shown in Table 1.

About 25% of the fatal crashes of 16- and 17-year-old drivers occur between 8 pm and 4 am on Friday or Saturday nights — a period in which alcohol is likely to have a great impact on the capability of drivers. About 18% of adult fatal crashes occur during this period. What is more, fatal crashes among teen-agers are likely to cause about two deaths for persons other than the driver, whereas if the driver is over 25, about one other person is likely to lose his/her life if the driver is involved in a fatal crash.[1,2]

Not all of the deaths occur in automobile crashes. Young people also use motorcycles, and we have already accumulated some very convincing evidence as to the life-saving potential of laws which require that all motorcyclists wear crash helmets. Between 1967 and 1969, 37 of the states enacted laws requiring motorcyclists to wear helmets. Part of the inducement was in order to qualify for safety-program-related federal highway funds. Figure 1 shows the steep decline in national fatality rates per 10,000 motorcycles registered between 1965 and 1975. These remarkable decreases in fatal risk occurred during a period of generally increasing mortality rates for young adults.

Figure 1 also shows the frightening increase in the rate as some states began to repeal or weaken the laws. The pressure for repeal came from a mistaken demand of cyclists for freedom to make their own decisions as to their use of helmets. In four states which repealed the compulsory motorcycle helmet laws, surveys showed that the proportions of cyclists wearing helmets declined from nearly 100% during the law's existence to about 50% after repeal. Since many motorcycle injuries lead to spinal cord and nervous system damage, it is not only the cyclists themselves who have to pay the huge and tragic toll for death and injury, but often their families and the community at large.[4]

Every reasonable person acknowledged that such laws were difficult to enforce, but in addition to the legal penalties, such laws also express a community judgment, and have an impact in defining "socially acceptable behavior".

The same concern about personal freedom led many states to lower the minimum age for the legal purchase of alcoholic beverages when Congress voted in 1970 to lower the minimum age for voting in federal elections from 21 to 18 years of age. Michigan lowered its minimum age for legal purchase of alcoholic beverages from 21 to 18 in January 1972.

Table 1
NUMBERS OF DRIVERS IN FATAL CRASHES BY AGE AND SEX AND RATES PER 10,000 LICENSED AND PER 100 MILLION MILES DRIVEN (U.S., 1978) (SELECTED AGES ONLY)[3]

	Ages						
	16	17	18	19	20—24	40—44	60—64
Males							
Fatal crash drivers	1,266	2,032	2,695	2,717	11,435	2,980	1,558
Per 10,000 licensed	11.9	12.5	14.7	14.0	11.5	5.2	3.2
Per 10^8 miles	48.9	24.6	19.9	12.6	8.3	3.1	3.3
Females							
Fatal crash drivers	329	465	547	549	2,218	613	425
Per 10,000 licensed	3.8	3.6	3.6	3.3	2.5	1.2	1.2
Per 10^8 miles	19.8	12.8	9.0	5.4	3.8	1.8	

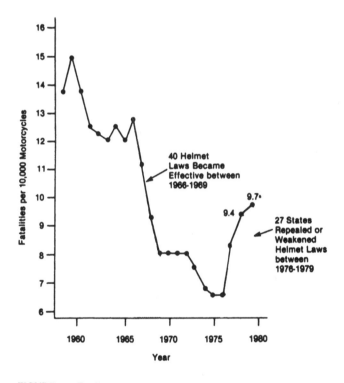

FIGURE 1.   Fatality rates by year in the U.S. for motorcyclists, showing the preventive impact of adoption of laws requiring motorcyclists to wear crash helmets which were adopted by 37 of the states between 1966 and 1969. After 1976, 27 states repealed or weakened these laws and an abrupt rise in fatalities per 10,000 motorcyclists occurred.

II. THE IMPACT OF LAW ENFORCEMENT ON DRIVING UNDER THE INFLUENCE OF ALCOHOL

The deterrent potential of law enforcement in dealing with the problems of the drinking driver has recently been reviewed,[5] based on experience in many different countries. It is clear from this review that the public debate over the twin issues of the likelihood of detection and the certainty and severity of sanctions has a great impact relative to the actual likelihood of detection and of penalty. Put another way, the changes in the law which have had the greatest impact have been those around which there was the greatest controversy, and hence the greatest publicity, and not necessarily those for which the actual change in the law or penalty was most substantial.[5] In nearly every case the demonstrable impact of the change in these laws diminished with the passage of time, presumably reflecting the nearly inevitable awareness that being caught and punished was rather improbable even if the law was broken and accident or injury followed. It is a surprise to realize that the countries with the most impressive evidence of a favorable impact of laws making driving with high blood alcohol levels a criminal offense are the very countries with low young adult mortality from motor vehicle crashes. Great Britain, The Netherlands, Sweden, and Norway have all strictly enforced drunk driving laws and relatively low overall young adult mortality rates. In contrast, such apparently similar countries with high young adult mortality rates, France, Australia, New Zealand, Canada, and the U.S., have less strict legislation and/or enforcement. Finland is an interesting exception, with unusually severe penalties for persons convicted of driving under the influence of alcohol, in that it has a low mortality rate from motor vehicle crashes of young adults. Thus, the deterrent effects of new legislation are partly dependent on the public belief as to the likelihood of arrest and punishment and tend to be transitory, but can make at least short term reductions in mortality of about 20%. This section has dealt with the prevention of driving after drinking for drivers of whatever age.

Up to now we have introduced three possible strategies for possible reduction of the toll of alcohol-related motor vehicle injury and death:

1. Deterrent effects of the law and its enforcement.
2. Required use of "packaging" or protective apparatus.
3. Education and changing of public fear or attitude. (For more comprehensive treatment of the possible accident prevention strategies, see Haddon[6].)

The use of crash helmets for motorcyclists is an example of a change in the law concerning protection from injury once a crash occurred. It tended to be selfenforcing in that cyclists without helmets were obvious. So this was a change in law *concerning packaging of the persons at risk* and it was *selfenforcing* and apparently quite effective in reducing the toll. Furthermore, the repeal of such laws apparently reversed the beneficial effects. The experience with laws making it illegal to drive after drinking were at least briefly effective, but were more so, the greater the publicity surrounding their enactment. While we haven't discussed it here, the effectiveness of educational programs alone has been low or absent.[7]

One possibility for reducing the toll of fatal teen-age motor vehicle crashes would be an increase in the minimum legal age for the purchase of alcoholic beverages from 18 to 21.

In 1978, Michigan voters passed a constitutional amendment put on the ballot by public petition to increase the minimal age for the legal purchase of alcoholic beverages from 18 to 21. The issue was highly controversial and involved the first and the last of these strategies.

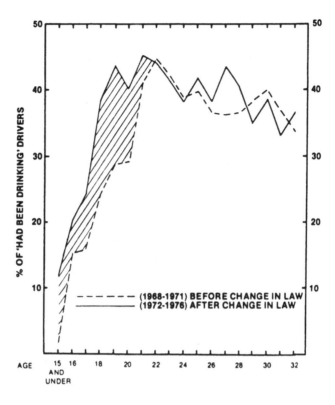

FIGURE 2. Age-specific proportions (%) of alcohol involvement of drivers in fatal accidents in Michigan during the period 1968—1971 when the minimum legal age for purchase of alcoholic beverages was 21 and during 1972—1976 when the age was 18. The shaded area reflects the impact of the lowering of the drinking age. (From University of Michigan Transportation Research Institute Report, March-April 1978.)

III. THE COMMUNITY

Michigan is a state with a population of about 9 million, and Detroit is its major metropolitan center. It is the home of the motor vehicle industry, and has a large area with relatively sparse population. It is justly proud of its university system, centered in Ann Arbor, about 40 miles west of Detroit. Located in Ann Arbor is the university's Highway Safety Research Institute (now renamed Transportation Research Institute) which was asked by the Office of Highway Safety Planning, Michigan Department of State Police to study the trend in fatal accidents involving alcohol between 1968 and 1976. Figure 2 shows the results of this study.

In the period 1968 to 1971 the approximately 500,000 18- to 20-year-old licensed drivers averaged 383 fatal accidents per year, of which 81 had, according to police records, involved alcohol. For the 1972 to 1976 period, the number of 18- to 20-year-old drivers had increased about 10%, and the average of fatal accidents was 410, but these an average of 158 fatal accidents involved alcohol. During the latter period, there was an overall drop in accident fatalities due to the enforcement of the 55 mph speed limit, but the number of fatal accidents involving drinking 18- to 20-year-olds had almost doubled. Table 2 shows some of the data for selected years, 1971, 1976, and 1977, for those who were younger than 18, those 18 to 20, those older than 24 years, and for all ages. Table 3 shows the same data in terms of rates per 100,000 licensed drivers. Those over 24 years of age in 1976 became 21 in 1972,

Table 2

DRIVERS INVOLVED IN FATAL AND INJURY ACCIDENTS, MICHIGAN 1971, 1976, AND 1977 BY SELECTED AGES AND WHETHER OR NOT THE DRIVER HAD BEEN DRINKING

Age	Type	1971	1976	1977
< 18	Fatal, total	184	204	202
	drinking	26	49	51
	Injury, total	13,825	15,485	16,014
	drinking	541	1,332	1,291
18—20	Fatal, total	369	439	492
	drinking	87	202	193
	Injury, total	24,680	29,515	30,487
	drinking	1,787	5,241	5,182
24 +	Fatal, total	1,780	1,529	1,581
	drinking	420	451	467
	Injury, total	108,230	111,407	113,943
	drinking	13,735	13,411	13,555
All ages	Fatal, total	2,788	2,603	2,672
	drinking	678	894	876
	Injury, total	173,986	184,871	190,224
	drinking	19,597	24,689	25,063

From Michigan State Police Data Center.

Table 3

RATES OF DRIVERS INVOLVED IN FATAL AND INJURY ACCIDENTS, MICHIGAN, 1971, 1976, AND 1977, BY SELECTED AGES AND WHETHER OR NOT THE DRIVER HAD BEEN DRINKING (RATES PER 100,000 LICENSED DRIVERS)

Age	Type	1971	1976	1977
< 18	Fatal, total	83	60	58
	drinking	11.7	14.4	14.8
	Injury, total	6202	4551	4660
	drinking	243	391	375
18—20	Fatal, total	76	82	93
	drinking	18.0	38.1	36.5
	Injury, total	5109	5561	5766
	drinking	370	987	980
24 +	Fatal, total	43	34	34
	drinking	10.0	10.0	10.2
	Injury, total	2591	2479	2487
	drinking	329	298	296
All ages	Fatal, total	51	43	43
	drinking	12.3	14.8	14.2
	Injury, total	3152	3055	3093
	drinking	355	408	407

From Michigan State Police Data Center.

the year the minimum age for legal purchase of alcoholic beverages was lowered from 21 to 18, and that is why the data for 21- to 24-year-old drivers are not tablulated.

Of the number of different statistical indicators of motor vehicle accident risk, none are entirely perfect. One can report the absolute numbers, but at different times and for drivers of different ages the proportion who drive is likely to differ; what is more, the average number of miles driven for drivers of different ages is also likely to differ. There are three common types of data reporting systems for fatal events.

One can report the number of deaths per 100,000 in the population; one can report the number of deaths per 100,000 licensed drivers, or one can report the numbers of deaths for each 100,000,000 miles of "exposure". In Table 1 we use two of these indexes; in Table 2, we give data in absolute numbers, and in Table 3 we show the rates per 100,000 licensed drivers. As can be seen from a simple arithmetic calculation, the numbers of licensed drivers increased sharply for the under 18-year-old population, and less so for the total pool of drivers and for the 18- to 20-year-old drivers. During all of the period the minimum age for licensure was 16 years of age.

The data from these tables tell a tragic story of what happened after the minimum age for the legal purchase of alcoholic beverages was lowered in 1972. The number of 18- to 20-year-old drivers involved in fatal accidents rose between 1971 and 1976-77, but the number of those who were drinking when involved in fatal accidents more than doubled. In the same time interval, those over 24 involved in fatal accidents actually decreased. For accidents involving injury, the 18- to 20-year-olds who had been drinking almost tripled, while for those over 24 the numbers practically didn't change. Those under 18 showed similar patterns, but somewhat less striking than that for the 18- to 20-year-old drivers. Data from Table 3 show that the rates per 100,000 licensed drivers for older persons involved in drinking-related accidents was almost the same in all three years, but it more than doubled for the 18- to 20-year-old drivers, and increased substantially for the under 18-year-old drivers. The injury rates for drinking drivers went up sharply for both groups of young people, but in those over 24, the rates dropped.

IV. THE AIMS OF COMMUNITY ACTION

Into this situation moved a church affiliated organization, the Michigan Council on Alcohol Problems, which initially tried to lobby the state legislature to raise the minimum age to 21, without success. After cooperating in the summer of 1976 with environmental groups in a petition campaign to ban throw-away bottles, it decided to use the petition procedure to put on the ballot a constitutional amendment to raise the legal age back to 21 years.

A. Methods
The Michigan Council on Alcohol Problems, under the leadership of Rev. Allen B. Rice II had to depend on church, PTA (Parent Teachers Association), and community volunteers, but between September 1977 and July 1978 it succeeded in getting over 325,000 signatures for its amendment. When it appeared that it might reach its goal of bringing the proposal to the voters, the Michigan Legislature suddenly decided to increase the age to 19. This coupled with a well-financed publicity campaign by the Michigan Licensed Beverage Association whose motto is "For Those Who Supervise and Control Drinking" seemed like formidable obstacles. The tavern owners' goal was to spend nearly 1 million dollars appealing for support for the civil rights of young people, with a vigorous effort to register 18- to 20-year-old voters.

The "Coalition for 21" as the supporters of the proposed amendment were known, campaigned on the slogan of "Saving 54 Lives a Year is Our Goal". They had a very small budget and virtually no newspaper support. The "Coalition" had, however, the support of the Michigan Medical Society, the Directors of the Departments of State Police and of Public Health, and most of the high school principals.

A group of high school students seeking to dramatize the difficulty of enforcing a minimal age for sale of alcohol held a press conference to claim that all of them, under 18 years of age, had no difficulty in buying alcoholic beverages. When challenged to name the stores which had violated the law, they refused; by doing so, their publicity backfired as this was cited as evidence that teen-agers were irresponsible.

Initial polls showed, however, that voters favored the amendment, and on election day they gave it a comfortable margin of 57.2%.

B. Results

Figures 3 and 4 show the rise and abrupt drop in police reported property damage crashes of 18- to 20-year-old drinking drivers and the rise and abrupt drop in police-reported 18- to 20-year-old drinking drivers involved in injury crashes. Injury involvement in alcohol-related crashes appeared to drop by 20%. Similar reductions in either alcohol-related or nonalcohol-related crashes among drivers aged 21 to 23 or 24 to 25 did not occur, but drivers aged 16 to 17 were involved in 17% fewer alcohol-related crashes and 9% fewer nonalcohol related crashes. The small reductions in other age groups over this period are due to other factors. These effects have been filtered out before arriving at these estimates. Effects other than those due to change in the minimum legal age are thought to be responsible for the continued downward trend shown for recent years in Figure 4. The impact of the change in minimum legal age continues to be manifest, however.[7]

Table 4 shows that there was an overall reduction of accident mortality for the age group 15 to 24 of about 20% and for suicide and homicide of about 15%, when the data are based on average rates for the three preceding years and the three subsequent years. For 5- to 14-year-old and 25- to 34-year-old populations there was only a 6 and 8% decrease respectively, in accident mortality rates over the same interval.

Specifically, the average mortality rate due to accidents in the three subsequent years (1979 to 1981) was 80.4% of the average in the preceding three years; in addition, the suicide and homicide rates were also down to 82.9 and 85.7% in 1979 to 1981 compared to 1976 to 1978, and the three rates combined were down to 81.4%.

We can use such data to make a variety of estimates of the number of lives that were not lost in the 1979 to 1981 period which would have had the mortality experience of the years 1976 to 1978 continued. Using the reduction to 81.4%, we could estimate that 18.6% was the proportional reduction from all three causes and 18.6% of 1577, the average number of deaths per year, gives 293 lives "saved" according to this definition. This is greater than the actual difference in the mean number of deaths, 246, because it takes into account the increase in the numbers of the population.

A more conservative estimate would correct this for the rate of decrease in accident deaths in 5- to 14-year-olds and 25- to 34-year-olds, 6 and 8%. Taking an average of 7% and subtracting this from the 18.6% would give a reduction of 11.6% of 183 deaths of 15- to 24-year-old young adults in Michigan per year.

This is over three times the 54 deaths per year the sponsors of the constitutional amendment used in their campaign for a higher minimum legal age for the purchase of alcoholic beverages.

No such argument can attribute all of the drop in mortality rates to the passage of the constitutional amendment alone, but the drop in Michigan was substantially greater than the drop in the U.S. as a whole. Since both specific and general effects can be documented, it is reasonable to conclude that the increase in the minimum age for purchase of alcohol in Michigan has resulted in saving of more than 100 lives of young adults per year.

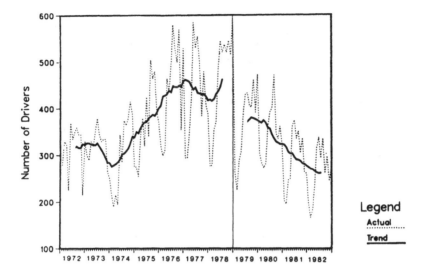

FIGURE 3. The numbers of drivers aged 18 to 20 involved in crashes with injury by month in Michigan and a 4-month moving average for the periods prior to and following the return to the minimum legal age for purchase of alcoholic beverages, January 1, 1979. (Compiled from Michigan police records by the University of Michigan Traffic Research Institute.)

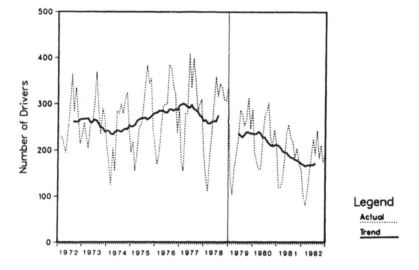

FIGURE 4. The numbers of 18- to 20-year-old male drivers involved in single vehicle night-time crashes with injury in Michigan by month. Four-month moving average is plotted separately for the period prior to and subsequent to the increase in the minimum legal age for purchase of alcoholic beverages. (Compiled from Michigan police records by the University of Michigan Traffic Research Institute.)

Table 4

MORTALITY AND MORTALITY RATES FOR 15- TO 24-YEAR-OLD POPULATIONS, MICHIGAN, 1976, 1977, AND 1978, COMPARED TO 1979, 1980, AND 1981, FOR ACCIDENTS, SUICIDE, AND HOMICIDE

	Numbers of Deaths							
Cause	1976	1977	1978	1976—1978 (mean)	1979	1980	1981	1979—81 (mean)
Accidents	1038	1030	1015	1028	938	911	722	857
Suicide	263	262	268	264	256	214	211	227
Homicide	325	267	262	285	239	274	229	247
Total	1626	1559	1545	1577	1433	1399	1162	1331
	Rates (per 100,000 per year)							
Accidents	62.0	60.6	59.2	60.6	54.5	50.7	40.9	48.7
Suicide	15.7	15.4	15.6	15.6	14.9	11.9	12.0	12.9
Homicide	19.4	15.7	15.3	16.8	13.9	15.3	13.9	14.1
Total	97.1	91.7	90.1	93.0	83.3	77.9	65.9	75.7

From Michigan State Department of Public Health, Office of Vital and Health Statistics.

V. IMPACT OF THE FINDINGS

The Michigan Licensed Beverage Association did not take the popular judgment as the final word, and sought in 1979 to obtain signatures to reverse the constitutional amendment which went into effect in the beginning of 1979, with another amendment lowering the age to 19. The University of Michigan Transportation Research Institute again was asked to see if the raising of the drinking age had any impact on fatalities, injuries, and property damage. The work was undertaken by Dr. Alexander Wagenaar as part of his academic work at the University of Michigan School of Public Health.

When it became apparent in the Spring of 1980 that the firm hired to gather signatures for the amendment to drop the age to 19 was going to have difficulty in obtaining enough signatures, the Michigan Legislature passed by a two thirds vote of both houses the resolution putting the amendment on the ballot. As the time for the Fall elections drew near, Dr. Wagenaar completed his report showing that indeed there had been a sharp drop in police-reported injury-producing, and property-damaging collisions among drinking 18- to 20-year-old drivers. Although there was a reduction in the fatalities, because of the small numbers, the change was not statistically significant. The state agency supporting the work sent the draft report out for review. Dr. Wagenaar had also submitted it as a paper to the Fall meeting of the American Public Health Association, which occurred shortly prior to the general election. When the results were presented at the meeting, there was a great deal of publicity, because of the high level of public interest in the issue. When the second vote was counted, the voters rejected the proposed amendment to drop the minimum legal age for the purchase of alcoholic beverages by an even greater margin, 62%.

VI. DISCUSSION

In fact, Michigan was but one of several states which changed its laws about the same time. Williams[8] and his colleagues at the Insurance Institute for Highway Safety in Washington, D.C., carried out an analysis of the impact of the changes of the laws in Illinois, Iowa, Maine, Massachusetts, Minnesota, Montana, New Hampshire, and Tennessee, using the experience of adjacent states for comparison. With such a large data set, it was possible to examine the effect of change in the law concerning the minimum legal age for purchase of alcoholic beverages on mortality.

Single vehicle night-time fatal crashes have been shown to involve alcohol in a high proportion of cases. Since it is considered possible that police reporting of alcohol involvement may be affected by the differences in the criteria used by the police in different jurisdictions, it is preferable to pool experience from a number of different locations using data on single vehicle night-time crashes. Williams et al. were able to show that there was an 18% net reduction in night-time fatal crashes involving drivers in states affected by the change in legal minimum drinking age, and a whopping 35% reduction in single vehicle night-time fatal crashes among this group. There was a smaller reduction in all fatal crashes, and all of these reductions were statistically significant, using a two-tailed test and criterion of 0.05.

As of January 1981, in the 14 states which raised the legal age for purchase of alcoholic beverages, it was estimated that there had been 380 fewer young drivers involved in night-time fatal crashes; since for each such driver two other fatalities on the average occurred, it is reasonable to assume that over 1000 lives have been saved. If all states had such laws, and the experience on which the analysis was based applied equally to them, an estimated 730 fewer younger drivers would be involved in night-time fatal crashes, and that much more death and injury from driving under the influence of alcohol could be avoided.

Unlike the benefits from crackdowns on driving under the influence of alcohol, which seem to be of brief duration, it seems from the Michigan experience that the raising of the minimal legal age for purchase of alcoholic beverages has a continuing beneficial effect.

Monitoring of the problem is being assisted by a Fatal Accident Reporting System operated by the National Highway Traffic Safety Administration. For an increasing fraction of the fatal accidents being reported, blood alcohol concentrations are also measured and reported. In 1981, 28% of the 4199 16- to 19-year-old drivers involved in *single vehicle fatal accidents* had positive blood alcohol tests compared to 14% of the 4591 teen-age drivers involved in *multiple vehicle fatal accidents*.[9]

A modest 5% decrease in fatalities among 15- to 19-year-old drivers in 1982 compared to 1981 was indicated by this data system, and mortality rates reported to the U.S. National Center for Health Statistics seem to confirm this trend.

Of the 101,703 drivers, passengers, and pedestrians involved in 1982 fatal motor vehicle accidents, 43,721 died. Of these, 41% were under 24 years of age, and 15% were 15 to 19 years of age. According to police reports, 36% of the deaths occurred in accidents in which at least one driver had been drinking; 34% of the *drivers involved in fatal accidents* were under 24 years of age.[10]

Thus, on a national level, there is still a long way to go to reduce the toll of alcohol-involved motor vehicle crash injury among young drivers. Michigan's experience shows one way to make a favorable impact on the problem.

Appendix
MORTALITY OF YOUNG ADULTS (AGE 15 TO 24) IN VARIOUS COUNTRIES BY SEX AND GROUPS OF CAUSES (PER 100,000 PER YEAR)

Country	GNP/capita	Sex	Total death	External causes	Disease	Motor vehicle
High Mortality — Developed Countries						
Switzerland	12,000	M	134.6	105.6	29.0	53.2
		F	43.4	27.1	16.3	14.1
U.S.	9,590	M	179.3	143.1	36.2	66.4
		F	62.5	37.9	24.6	19.7
Federal Republic of	9,580	M	147.4	112.6	34.8	72.3
Germany		F	56.6	32.3	24.3	13.8
Canada	9,180	M	174.9	147.0	27.9	79.0
		F	56.1	37.2	18.9	21.6
Belgium	9,080	M	134.3	100.0	34.3	69.4
		F	51.2	28.1	23.1	16.7
France	8,260	M	148.7	112.9	35.8	59.7
		F	56.7	32.4	24.3	17.6
Australia	7,990	M	162.4	133.0	29.4	93.0
		F	53.3	33.1	20.2	22.2
Austria	7,030	M	175.9	144.1	33.9	90.8
		F	53.8	33.1	20.7	21.0
New Zealand	4,790	M	165.5	131.6	33.9	85.7
		F	58.2	31.5	26.7	21.9
Finland	6,820	M	149.4	118.9	30.5	39.7
		F	46.8	24.5	22.3	11.1
Low Mortality — Developed countries						
Sweden	10,210	M	99.1	75.8	23.2	34.5
		F	41.8	26.0	15.8	10.7
Denmark	9,920	M	103.6	78.1	25.5	46.8
		F	38.2	21.8	16.4	11.2
Norway	9,510	M	107.5	78.4	29.1	35.7
		F	34.5	17.4	17.1	7.5
Netherlands	8,410	M	95.7	64.4	31.3	47.8
		F	38.1	19.6	18.5	13.6
Japan	7,280	M	95.8	64.9	30.9	30.4
		F	43.2	19.3	23.9	4.1
England & Wales	5,030	M	92.2	61.3	30.9	22.8
		F	39.1	17.0	22.1	8.0
Some Mediterranean Countries[a]						
Italy	3,850	M	103.4	70.9	32.5	45.9
		F	41.6	16.3	25.3	10.5
Israel	3,500	M	120.3	82.0	38.3	24.8
		F	46.4	21.3	25.1	10.7
Spain	3,470	M	99.7	60.2	39.5	28.1
		F	39.7	13.4	26.3	7.4
Greece	3,250	M	94.3	61.7	32.6	33.9
		F	40.3	15.3	25.0	7.4
Yugoslavia	2,380	M	109.8	71.7	38.1	na[b]
		F	52.4	20.5	31.9	na
Egypt	390	M	282.2	103.4	179.4	3.7
		F	175.0	41.8	133.2	0.4

Appendix (continued)
MORTALITY OF YOUNG ADULTS (AGE 15 TO 24) IN VARIOUS COUNTRIES BY SEX AND GROUPS OF CAUSES (PER 100,000 PER YEAR)

[a] The group of Mediterranean countries experience reflects the more modest impact of motor vehicles in countries whose affluence has not yet allowed many of its young adults to afford to risk death at the wheel. The data for Egypt, whose capital city, Cairo, must have the most automobile-choked streets in the world, reflects the high disease rates for young adults, but not a high rate of motor vehicle-related mortality.

[b] na, Not available. The data for countries with centrally planned economies do not separate out mortality from motor vehicle accidents.

From: World Health Organization, Health Statistics Annual, Geneva. The mortality data are averages for years 1972 to 1978 if available. The Gross National Product data are from the World Bank, and reflect the status as of 1975 in most cases.

REFERENCES

1. *Status Report,* The Insurance Institute for Highway Safety, Washington, D.C., 1983.
2. **Robertson, L. S.,** *Injuries,* D. C. Heath & Co., Lexington, Mass., 1983, 57.
3. **Robertson, L. S.,** *Injuries,* D. C. Heath & Co., Lexington, Mass., 1983, 57.
4. **Ross, H. L.,** *Deterring the Drinking Driver: Legal Policy and Social Control,* D. C. Heath & Co., Lexington, Mass., 1983, 18.
5. **Ross, H. L.,** *Deterring the Drinking Driver: Legal Policy and Social Control,* D. C. Heath & Co., Lexington, Mass., 1983, 99.
6. **Haddon, W., Jr.,** Options for the prevention of motor vehicle crash injury, *Isr. J. Med. Sci.,* 16, 45, 1980.
7. **Robertson, L. S.,** *Injuries,* D. C. Heath & Co., Lexington, Mass., 1983, 91.
8. **Williams, A. F., Zador, P. I., Harris, S. S., and Karpf, R. F.,** The effect of raising the legal minimum drinking age on involvement in fatal crashes, *J. Legal Studies,* 12, 169, 1983.
9. U.S. Department of Health and Human Services, Public Health Service, Morbidity and Mortality Weekly Report, Washington, D.C., July 8, 1983.
10. U.S. Department of Health and Human Services, Public Health Service, Morbidity and Mortality Weekly Report, Washington, D.C., December 16, 1983.

Chapter 18

THE LONG VIEW: LESSONS FROM EXPERIENCE

John R. Goldsmith

In four of the last five chapters, the focus was on short term effects and their consequences even though the problems had important long term implications. We really have no information at all on the long term implications of occasional interference with transport of oxygen in babies during the first 3 months of life. We cannot ethically experiment with them, and the types of effects which were documented by Needleman et al. in school children exposed to lead would be nearly impossible to detect in experimental animals. We are therefore not likely to get this information. It is far more prudent to plan a course of action which neither requires us to prove that mental and emotional development are affected nor that cancer increases as a result of ingestion of drinking water with elevated levels of nitrate. We must depend on the evidence of short term effects, in this case elevated methemoglobin levels as a guide for prevention of long term problems, fully aware that the facts which will confirm both the need and the benefit from such protective measures may never be forthcoming.

Carbon monoxide is unique as a major air pollutant in that its presence in elevated amounts produces no sensory clue; not only do people never complain of carbon monoxide exposures as a community pollutant (in contrast to the headache and drowsiness it may produce in occupational exposures), but despite the statistical evidence that it is associated with deaths from cardiovascular causes, no single death can be or has been attributed to community carbon monoxide exposures. Nevertheless, the evidence has been considered sufficient so that a substantial effort and expenditure has been made to reduce the community exposures and exposure levels have been reduced as a result. The reasons for this are the strength of the evidence from experimental studies and the extent to which community study data agree with what would be predicted on the basis of experimental findings. Although the case study in Chapter 14 deals with a short term effect, other laboratory and epidemiological evidence suggests strongly that carbon monoxide speeds up the development of arteriosclerosis. Difficult as it is to establish the evidence of a short term effect, that evidence is still more convincing than that relative to long-term effects at the dose levels likely to be encountered as a result of community pollution. For such long term effects, the most powerful evidence is from comparing cigarette smokers and nonsmokers, but there are other agents in cigarette smoke than carbon monoxide. Disentangling the effects of a single agent under these circumstances is a difficult task. It is therefore more prudent to use the evidence of short term effects on survival of hospitalized patients with myocardial infarction as the guide for our preventive efforts than to wait for equally compelling evidence as to long term effects.

In the studies of learning disabilities possibly associated with lead exposures in early years of life, we have an example of a problem with a medium term (2 to 4 years) latency period between exposure and manifestation of the learning difficulty, but a very long subsequent period of handicap, which extends throughout life, as a likely result of poor classroom performance, and the anticipated sequence of fewer years of education, and lower learning and earning capacity. If community lead exposures are responsible for even a small increase in the number of persons who need institutional care as a result of intellectual or motor impairment, then the social and personal costs are very high indeed. The evidence presented in this chapter is strong enough to justify a vigorous preventive program. The Administrator of the U.S. Environmental Protection Agency in January of 1984 announced his support for complete removal of lead from motor vehicle fuel, and the evidence presented in Chapter 15 was the most persuasive.

The efforts of Dr. Rokaw and Professor Detels to document the long term health effects of photochemical pollution represent the only study in this section with the principal objective of identifying a long term effect. It is a long, costly and complicated effort, and the work is still incomplete. So far it has not indicated a need for more restrictive pollution controls (that is to lower levels) than are indicated for prevention of acute effects. We must not forget that those who resist control programs will keep emphasizing that the symptoms and lung function impairment which are the principal current scientific bases for the regulations are transient and reversible problems. So with such long term studies we may not identify further controls that are needed, but we may strengthen the effort to achieve the air quality goals reflected in the present standards.

Alcohol and substance abuse are immense social, health, and economic problems. From the point of view of excess mortality, the most obvious impact is on deaths of 15 to 25 year olds. Among this age group, there is likely to be a confusion of attitudes toward personal perogatives and social obligations. There is also unfortunately no organized political representation for these young adults. As a result, little attention has been paid by epidemiologists to the striking changes in mortality on a year to year and state by state basis. The Michigan case study provides evidence that laws affecting drinking behavior can have a favorable impact on mortality in this group. Since the economic value to society reflecting the investment in education, nurturance, and the potential future earnings is at its maximum in the 15- to 25-year-old group, the long term consequences of steps to reduce the mortality of young adults are very great. We are still searching for reasonable steps to take, but the full development of preventive programs will need more input from young adults themselves, and possibly a greater level of political influence either by them or on their behalf. Some part of the present weakness in recognition and response to health, social and environmental problems of young adults is the lack of an adequate information base. Young as well as more mature people need facts as a basis for actions they may want to take in order to protect their health and their lives. The relevant facts include not only data on exposures and effects, but inferences as to risks, and facts as to how community decision processes work and how they may be made more responsive.

The experiences of the authors of this book have taught us lessons, which we earnestly wish to share as widely as possible, and particularly with students who must look forward to a long period of consuming, producing, and hopefully preventing pollution.

Part E

Chapter 19

REPRODUCTIVE EFFECTS OF DBCP: IT TAKES TWO TO MAKE A BABY

John R. Goldsmith and Gad Potashnik

TABLE OF CONTENTS

I. THE PROBLEM

There are many new chemicals tested for their effectiveness — as pesticides, as growth promotors, as food preservatives, as cosmetics — but in addition to their effectiveness, under present laws, nearly all must be tested for health risks. Some of the testing is thorough and careful; some of it is superficial and sloppy. Some of the testing is openly published; some was or is kept as a "trade secret" for the advantage of one firm in the highly competitive chemical industry.

This is the story of one chemical, 1,2-dibromo-3-chloropropane, known as DBCP, which was thoroughly and carefully tested. The results of the tests were published by Torkelson and his colleagues in a widely read journal,[1] but several major chemical firms went ahead and produced and marketed it without paying attention to one of the risks which the testing procedures had clearly indicated! The risk which later became a reality was of male sterility.

I recall a wine tasting party in San Francisco in the late 1950s or early 1960s when a colleague asked if I thought there would be market for a male contraceptive. When I replied affirmatively, he said: "Well, we have found one, but it has some other features which won't make it very popular and anyhow it takes some time to be effective." He was one of the authors of the article[1] describing the testicular toxicity of DBCP.

It was an unusually effective agent for killing nematodes in the soil as well, because a number of the largest chemical companies began producing and selling it in large amounts. (DBCP is called a nematocide, for its lethal effect on nematodes, a primitive worm living in the soil which damages crops.) At least one of the companies monitored its production worker's liver function, in the mistaken impression that like vinyl chloride, the initial target organ of DBCP would be the liver.

At that time concern about possible reproductive effects of chemicals was concentrated on the hazards for exposures of women who may have been in the early weeks of pregnancy.

In 1977, five production workers at Occidental Chemical Co. in California were comparing notes about their families, when it became apparent to them that, although they were using no contraceptives, none of their wives had become pregnant during the last few years. They went to their union representative for the Oil Chemical and Atomic Workers International, a union with strongly developed concerns for its members' health. The union sent them for examination to specialists, and they were found to have no spermatozoa in their semen — they were sterile. Other men with similar exposures were also examined and those with the longest periods of exposure were found to have no spermatozoa, and those with shorter periods had reduced numbers. The report of Torkelson et al. was reread with some embarrassment and the doctors in Berkeley reported their findings in the *Lancet*.[2]

The men affected did not appear to have any other interference with sexual function or performance despite some secondary disturbances in hormonal levels. After excluding the men who had had a vasectomy and those with intermediate sperm counts, two groups of 11 men were compared, one group with less than 1 million sperm per milliliter of semen, and another group with normal sperm counts, over 40 million/mℓ. These two groups differed most strikingly in the duration of their DBCP exposures. It averaged 8 years in the former group and 0.8 years on the latter group.

II. THE COMMUNITY

DBCP was manufactured, among other places, in an Israeli plant in Beer Sheva, and at about the same time that the Occidental workers became aware of their infertility problem, the Israeli workers were being evaluated and the results published in the first of several reports by Potashnik et al.[3]

Two unique features of the Israeli community of workers were that none had vasectomies, and nearly all wanted to have large families, so that potential reproductive events per worker were relatively frequent. In addition, the health care system in the Negev region of Israel where the plant is located, has a single full coverage health care plan, so that the workers and their families continued to be followed by a single medical team and there was virtually no loss to follow-up. Some of the workers were Beduin and some Jewish, but they had similar work assignments and identical health care.

An additional population of Jewish workers who were applying the material in the cultivation of bananas in the Jordan Valley, were studied in a cross-sectional study to see if there was any evidence of unfavorable reproductive outcomes in their families. In California, Glass and his colleagues[4] had reported that sperm count depression occurred in farm workers using the material over a sufficient number of seasons of application.

III. HYPOTHESES

We wished to test the hypothesis that men with low sperm counts (oligospermia) or no sperm at all (azospermia) as a result of DBCP exposures could recover the ability to have normal children.

We also wished to test whether unfavorable reproductive outcomes had occurred in the families of men who were using the material as applicators.

IV. METHODS

In 1977, 30 production workers were initially evaluated, and all have been followed for at least 5 years after production ceased. Of these, factory records indicated that 26 men had been exposed and for how long. The remaining four were employed in maintenance or administrative posts, and were not thought to have production exposures. These four men were evaluated at their own request. All four had normal sperm counts and are shown as "unexposed" in the tables following. Standardized procedures were used for semen collection and analysis and for obtaining and classifying exposure histories. Men characterized as "azospermic" had no sperm in repeated semen samples obtained in 1977 when production ceased. Men were characterized as "oligospermic" who had fewer than 20 million sperm per milliliter of semen on repeated tests in 1977.

Detailed reproductive histories were obtained from the wives of all of the men and confirmed from the records available in the Soroka Medical Center and the pre- and postnatal clinics of the health department.

Hormone measurements were made in October 1977 and October 1981 to determine the levels in blood plasma of the hormones which normally regulate testicular function. (Three hormones are involved, follicle stimulating hormone [FSH] and luteinizing hormone [LH] regulate testicular function. These hormones are the type that are excreted by the pituitary gland in increasing amounts into the plasma when the cells which normally respond to them are damaged. The third, testosterone, a hormone secreted by the normal testicular cells could be expected to be low if the cells which secrete it are damaged. Potashnik et al.[3] had shown FSH and LH to be significantly elevated and testosterone to be slightly and nonsignificantly depressed in their 1978 report.)

In the follow-up study, we assume that any baby born a year or more after the start of exposure was a child of an "exposed" pregnancy. Such babies may have been conceived at times when testicular function was diminishing or as it was recovering. Since our detailed information does not indicate any differences, once pregnancy was recognized, between the reproductive outcomes before and after cessation of production in 1977, we treat these as a single group.

In 1979, DBCP was still being used as a nematocide in the Jordan Valley banana plantations, since no equally effective substitute was known. A list of applicators was obtained from the banana cooperative and the men were interviewed concerning their schedule of work in bananas, the chemicals they had used and the reproductive histories of their families. The Ministry of Labor had scheduled sperm counts monitoring of the men, and so the field survey did not include these tests. Unfortunately no data on such counts were ever obtained. Reproductive histories were validated by checking hospital records at the local hospitals.[7]

V. RESULTS

The relationship of exposure and sperm count depression was striking. Thirteen men who were azospermic in 1977 had estimated exposure times of between 100 and 6727 hr. Eight men who were oligospermic in 1977 had exposure times of between 34 and 95 hr. Five normospermic men had exposure times of between 10 and 60 hr.

Four of the thirteen azospermic men recovered, in the sense that sperm were regularly produced, and of these two fathered normal children, while the wife of a third man had a hysterectomy. Three of the four who recovered were the three youngest (age 20 to 22) men of the thirteen, and the fourth had a relatively short exposure time, 115 hr.

Five out of seven oligospermic men had increases in sperm count, of which two reached the normal range; both of these fathered children.

The recovery of sperm counts often did not occur until several years had passed. This is shown in Figure 1.

Table 1 shows the reproductive outcome of "unexposed" and "exposed" pregnancies. The unexposed pregnancies include those of the wives of unexposed men as well as the pregnancies of the wives of exposed men before exposure started (pre-exposed). Similarly the exposed pregnancies are of the wives of exposed men who were either azospermic, oligospermic, or normospermic in 1977, and for which conception occurred at any time more than 3 months after the beginning of exposures.

Of 67 unexposed pregnancies, 16 terminated in abortion, half of which were spontaneous. As a result of these pregnancies, 27 male and 24 female children were born. Of the 22 exposed pregnancies, 6 were among the wives of azospermic men, 9 among wives of oligospermic men, and 7 among the wives of normospermic men. Of the 15 exposed pregnancies among wives of men with impaired spermatogenesis (that is with azospermia or oligospermia) in 1977, there were 2 induced and 1 spontaneous abortions. As a result of these pregnancies, two male and ten female children were born. The reasons for the induced abortions were unrelated to exposures.

Table 2 shows the health status of babies born to DBCP workers, by exposure status of the pregnancy. With the exception of one baby with a hemangioma, no unfavorable outcomes occurred among the 12 babies born to wives of men with spermatogenic impairment.

Among the unexposed pregnancies, two babies with congenital defects and one infant death occurred in a Beduin family in which the parents were first cousins. One other infant death was in a Beduin family. Pre-exposure pregnancies date back to periods when sanitation and prenatal care were poor, especially among the Beduin, and infant mortality rates were therefore relatively higher than in recent years. Four of the females and one male baby born of exposed pregnancies were born to Beduin. Four azospermic, four oligospermic, and one normospermic men were Beduin.

Table 3 shows the principal results of the study of applicators. No unusual frequency of problems was observed among the babies born to wives of men after exposures started. What was unexpected was the sharp rise in frequency of spontaneous abortion among the wives of men after, as opposed to before, exposure. Although not adjusted for age of the mother and parental smoking, both of which can increase the frequency of spontaneous abortion (see Chapter 20), the results are statistically significant at the $p < 0.05$ level. There is a possibility that such a result could be due to a tendency for the men to selectively recall

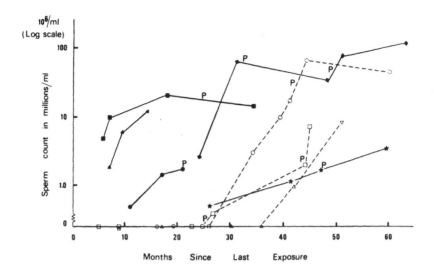

FIGURE 1. Recovery of spermatogenesis among DBCP-exposed production workers by months since last exposure. The solid symbols and lines refer to workers who were oligo-spermic when exposure ceased. The dotted lines and open symbols are those who were azospermic when exposure terminated. The symbol "P" indicates a recognized pregnancy.

Table 1
DBCP EXPOSURE AND REPRODUCTIVE OUTCOME

Spontaneous Abortion

Exposure category	Total pregnancies	Abortions	
		Induced	Spontaneous
Pre-exposed or unexposed	67	8	8
Exposed			
Azospermic (100 + hr)	6	2	0
Oligospermic (35—95 hr)	9	0	1
Normospermic (< 60 hr)	7	0	2

Sex Ratio

Exposure category	Liveborn	Males	Females	Males/total
Pre-exposed or unexposed	51	27	24	0.53
Exposed				
Azospermic (100 + hr)	4	0	4	0.00
Oligospermic (35—95 hr)	8	2	6	0.25
Normospermic (< 60 hr)	5	4	1	0.80
Azospermic and oligospermic	12	2	10	0.17[a]

[a] Significantly different from the binomial expectation based on Israeli population q of 0.4855 (p = 0.015).

and report spontaneous abortion after being exposed or because of the shorter time interval between exposed and unexposed periods. In checking the reports against hospital records, we found that in fact, there was underreporting of abortions during the exposure period.

Table 2
HEALTH STATUS OF BABIES BORN TO DBCP WORKERS

Health indicator	Pre-exposed or unexposed	Exposed
Number and sex	M 27 F 24	M 6 F 11
Birth defects	5 — Patent ductus, horseshoe kidney[a] (fatal) Hemangioma Hypospadias Umbilical hernia, septal defect Hemangioma[a]	1 Hemangioma
Infant mortality	4	0
Prematurity	3	1 Normospermic father
Birth weight (grams)	M 3379 ± 725 F 3393 ± 459	3383 ± 730 3407 ± 521
Morbidity	1 — Fatal pneumonia at 2 years	None — normal development

[a] These are both children of first cousin marriages. Another child of this couple was premature and died.

[b] Birth weights only available for babies born in hospital. Data refer to 33 unexposed pregnancies out of 51; 14 exposed pregnancies out of 17 have birth weight.

Table 3
SPONTANEOUS ABORTIONS AND RATES AMONG WIVES OF 66 MARRIED APPLICATORS PRIOR TO AND SUBSEQUENT TO FIRST EXPOSURES TO DBCP

	Prior to exposure	Subsequently
Number of pregnancies	76	121
Spontaneous abortions	5	24
Live births	71	97
Percent of abortions	6.6	19.8

From Kharazzi, M., Potashnik, G., and Goldsmith, J. R., *Isr. J. Med. Sci.*, 16, 403, 1980. With permission.)

VI. INTERPRETATION OF THESE RESULTS

Concerning the hypotheses to be tested, recovery from DBCP exposures is possible if the duration of exposure is relatively short. On the basis of this small sample and follow-up, those who recover father normal children. There is suggestive evidence that there is an increase in the frequency of spontaneous abortion in the wives of men who use DBCP; there is no evidence that with exposures intense enough to lead to azospermia or oligospermia, there is an increase in spontaneous abortion. The highest rate of spontaneous abortion among wives of exposed production workers was among those so lightly exposed as to remain normospermic, 2 out of 7 pregnancies, but such an effect could easily have occurred by chance.

For ten female babies out of twelve to be born to wives of men with evidence of impaired spermatogenesis is well outside the sex ratio to be expected on the basis of chance.

VII. IMPLICATIONS OF THESE FINDINGS

A. Why was the evidence for testicular toxicity in experimental animals ignored when a decision was made to produce and use this chemical?

Scientists for decades had known that one useful test for the mutagenic activity of chemicals (mutagenic activity is activity which changes the genetic structure of cells while allowing them to continue to reproduce) is to expose a male animal, mate it, and count the number of offspring. There had nevertheless been little previous interest and almost no reports of studies looking for such effects in exposed humans. Secondly, in order to obtain a semen sample from an exposed man, masturbation is commonly suggested, but there are complex feelings about masturbation, and there may well have been a reluctance to request such samples even by professional workers.

Finally, the exposed workers did not have any complaints relative to spermatogenic impairment to suggest that there was a problem to investigate. There was no previous experience with spermatotoxic agents to provide guidance.

As with many other chemical agents there were financial and technical inducements to encourage its manufacture and use.

There have been three consequences of the oversight involved in the DBCP experience. First, occupational health personnel have realized that exposures of men are relevant to the reproductive outcomes of their wives, and many studies have subsequently been directed to such problems. Monitoring of sperm counts using samples obtained by masturbation is no longer considered out of bounds for men who are at risk of potential spermatotoxic or mutagenic agents. Secondly, there is a great deal more vigilance in reviewing the toxicological data for new chemicals to be used in agriculture and for other purposes. Thirdly, the possible links between spermatotoxicity and mutagenicity are being actively explored. In recent years the availability of bacterial tests for mutagenicity have permitted a rapid advance in screening chemicals. DBCP has been found to be a potent mutagen for bacteria. One reason for this interest is the extensive overlap of agents which are mutagenic and carcinogenic (that is capable of producing cancer). DBCP has been shown also to be carcinogenic. So there is a potential link between protection from risk of cancer and evidence of reproductive effects of new agents. It is apparent that human excess cancer takes years or decades to mainifest itself, whereas reproductive effects may be identified in a much shorter period of time.

B. Two unanticipated findings were that there appeared to be excess numbers of abortions among the wives of men with low intensity and intermittent exposures (applicators), and that wives of more heavily exposed men (production workers) had an unusually high proportion of female children. What implications do these findings have for protection of workers and the general community?

First, we have in the past tended to look for one relatively uncommon effect of occupational exposures, namely congenital defects, on the assumption that either a mutagenic mechanism or some toxic effect on the developing fetus (called a teratogenic effect) was likely to be involved. In the case of DBCP, birth defects have not yet been found to be increased, but in addition to the infertility, there were two other types of unfavorable outcome. We are thus, in the future, more often going to look for *multiple as opposed to single unfavorable pregnancy outcomes of chemical hazards*. This becomes that much more important because sex ratio changes and increased frequency of spontaneous abortion may be statistically more readily detectable as a result of the same small relative changes than are rarer events like major malformations. Monitoring of the multiple reproductive outcomes of industrial workers now becomes more important than looking for a single unfavorable type of relatively rare outcome. A recent study of the histories of reproductive outcomes of a group of men in another plant which was manufacturing DBCP prior to 1977 demonstrates that *the effect could have been detected by monitoring of birth records and comparing them to expected reproductive rates at a time several years earlier than the spermatotoxic was in fact noted.*

C. What are the regulatory and legal implications of this experience?

Manufacture of DBCP in the U.S., was terminated in 1977, but its use on pineapple in Hawaii was permitted so long as the exposures were less than 1 ppb. Sperm counts among pineapple workers were monitored and found to remain unaltered. Reproductive histories of these men's families have not been studied.

Lawsuits by workers against their employers have been filed and are still in the courts. The possibility of recovery of normal spermatogenic function is relevant to these law suits.

DBCP has been found in well waters in locations in which it has been heavily used. Since it is animal carcinogen and mutagen this sets off alarms. However the concentrations are very low and the possibility of risk is not clear.

D. What are the biological consequences of the finding of a possible predominance of femal children in the offspring of exposed workers?

One group of scientists has reported a structural abnormality in the Y-chromosome in the spermatozoa of exposed workers.[5] We know that the Y-chromosome is the one which determines whether the offspring will be male. These scientists used a fluorescent dye which is known to attach to and stain the Y chromosome. They stained and examined for fluorescence, semen samples from workers exposed to DBCP and from other men, not exposed to known spermatotoxic agents. In both groups, they found approximately the same proportion of stained spermatozoa, so there appeared to be no difference in the proportion of male (stained) and female (unstained) spermatozoa associated with DBCP exposures. But they found something else. The DBCP worker's sperm had a higher proportion of sperm with two staining points, and this difference was statistically significant. They inferred that this reflected what is called "non-dysjunction of the Y-chromosome". By this was meant that when the sperm cells were being formed in the testes, a larger proportion of them in the DBCP workers failed to separate the two Y-chromosomes, so an increased proportion of the Y-chromosomes were faulty and probably could not fertilize an ovum. Such a method only permits the detection of one possible fault in the production and replication of the Y-chromosomes. Possibly there are other defects in the Y-chromosome or other chromosomes as well.

Although there is much interest in various environmental exposures which may affect the human sex ratio, this report is the first suggesting an impairment for a specific chemical agent. Previous reports had found 13 females out of 18 live births for Danish female anesthetists and 14 female births out of 21 first babies born after start of exposures for the wives of Danish male anesthetists (see Chapter 20). No confirmation of such an effect has been shown, although increased frequencies of congenital abnormalities and of spontaneous abortion are reported for both female anesthetists and wives of male anesthetists (see Chapter 20). A small but not statistically significant increase in the frequency of doubly staining spermatozoa was found in the semen of men exposed to carbaryl (an insecticide with the chemical name of 1-naphthyl-methyl carbamate, and the trade name of Sevin®); such men had significant increases in the proportion of abnormally shaped sperm, but no significant differences in sperm counts. However 7 out of 48 (14.6%) exposed men were oligospermic compared to 2 out of 34 (5.9%) of the unexposed men. Carbaryl also produces toxic effects in the testes of experimental animals, according to scientific work in the Soviet Union. Effects on reproductive outcome in the families of production workers were looked for but the findings have not yet been reported.[6]

Other reports show a predilection for female births among both men and women with lymphoma. During epidemics of Hepatitis B there has been an increase in births of males.

Much more work with experimental animals and with high risk populations of men and women are indicated in order to put the problem in such a perspective that we can confidently protect people from reproductive risks.

In the meantime, occupational epidemiologists have learned that it takes two parents to make a baby; healthy babies are more likely in families in which neither parent is exposed to toxic agents at critical periods of reproductive life.

REFERENCES

1. **Torkelson, T. R., Sadek, S. E., Rowe, V. K., Kodama, J. K., Anderson, H. H., Lozuvam, G. S., and Hine, C. H.,** Toxicological investigations of 1,2-dibromochloropropane, *J. Toxicol. Appl. Pharmacol.,* 3, 549, 1961.
2. **Whorton, D., Krauss, R. M., Marshall, S., and Milby, T. H.,** Infertility in male pesticide workers, *Lancet, ii,* 1259, 1977.
3. **Potashnik, G., Ben-Aderet, N., Israeli, R., Yanai-Inbar, I., and Sober, I.,** Suppressive effect of 1,2-dibrom-3-chloropropane on human spermatogenesis, *Fertil. Steril.,* 30, 444, 1978.
4. **Glass, R. I., Lyness, R. N., Mengle, D. C., Powell, K. E., et al.,** Sperm count depression in pesticide applicators exposed to dibromochloropropane, *Am. J. Epidemiol.,* 109, 346, 1979.
5. **Kapp, R. W., Jr., Picciano, D. J., and Jacobson, C. B.,** Y-chromosomal nondisjunction in dibromochloropropane exposed workmen, *Mutat. Res.,* 64, 47, 1979.
6. **Wyrobek, A. J., Watchmaker, G., Gordon, L., Wong, K., Moore, D., II, and Whorton, D.,** Sperm shape abnormalities in carbaryl-exposed employees, *Environ. Health Perspect.,* 40, 255, 1981.
7. **Kharrazi, M., Potashnik, G., and Goldsmith, J. R.,** Reproductive effects of dibromochloropropane, *Isr. J. Med. Sci.,* 16, 403, 1980.

Chapter 20

REPRODUCTIVE OUTCOMES ASSOCIATED WITH OCCUPATIONAL EXPOSURES TO ANESTHETIC GASES

John R. Goldsmith

TABLE OF CONTENTS

I. THE PROBLEM

Inhaled anesthetic gases are a group of the miracle drugs which has spared humanity pain and made possible feats of surgery which offer hope of cure for the otherwise incurable or deformed.

Ether, chloroform, and nitrous oxide, now joined by ethylene, cyclopropane, halothane, and many others represent an incomparable boon to mankind and womankind. From dental surgery or delivery room to the huge complexes in which teams now transplant organs, anesthetic gases are relied on to prevent pain and suffering while permitting modern technology to develop and be applied.

So important have such uses become that several occupational specialties are now recognized to administer such anesthetics. Anesthetists are medically trained specialists, who usually head up teams of anesthetists, nurse-anesthetists, anesthesiology technicians, nurses, technicians and orderlies, who together with surgical teams of similar categories are exposed daily to levels of anesthetic gases less than sufficient to cause anesthesia, but with other possible effects. The reason they are exposed is obvious if one realizes that most of the anesthetic gases are excreted in unaltered form by the body in the exhaled breath. The work of administering and regulating the dosage of anesthetic gases requires that the anesthetist keep his or her head near the patients's expiratory air stream in order to monitor pulse, eye reflexes, and breathing. It seems inevitable that such work is associated with regular but low level (relative to anesthetic levels) exposures to anesthetic gases.

That such exposures might be an occupational hazard had not been considered in advance; the health service occupations are ironically about the last to consider that they themselves may be at occupational risk from chemical agents, or for that matter from radiation, infectious agents, psychological stress, or musculoskeletal strain. By contrast, a very great deal of information had been assembled with respect to briefer higher level exposures of patients. Evidence as to the toxicity of chloroform for the livers of anesthesized patients, for example, had led to its being discarded.

In 1967 Vaisman[1] reported from the Soviet Union that 15 of 31 pregnancies among 110 female anesthetists resulted in spontaneous abortion. In 1970 Askrog and Harvald[2] reported on a questionnaire survey of Danish nurses working in anesthetic departments of hospitals (N = 578) and 174 female and male anesthetists. Satisfactory questionnaires from 570 people were returned documenting the outcomes of 212 pregnancies occurring prior to exposure and 392 during exposure in the anesthetic department.

Table 1 shows the reproductive outcomes according to whether the pregnancy was prior to exposure, during the first, second, third, or subsequent year of exposure, and according to whether the pregnancy resulted in spontaneous abortion, or the baby had a malformation. Since age is associated with an increase in the likelihood that a pregnancy will lead to spontaneous abortion, the mean ages of the women in each group are also shown. *There is an apparent increase in spontaneous abortion associated with both maternal and paternal exposures.*

Table 2 shows that exposure tends to lead to a disproportionate number of female babies. This apparent effect is most pronounced among the nurse anesthetists during the second year of exposure and wives of anesthetists during the first year of exposure. The nature of the exposures is not clearly defined, nor is its magnitude. Nevertheless, this work led to an awareness that such problems required study.

II. THE COMMUNITY

In occupational health studies, the group at risk may be working in a single factory or be in many scattered units. For the purposes of the chapter, we are going to define the

Table 1
REPRODUCTIVE OUTCOMES IN DANISH COUPLES, ONE OF WHOM WAS EXPOSED TO ANESTHETIC GASES[2]

Group	Pre-exposure pregnancies	1st year	2nd year	3rd year	3+ years	All exposed
Nurse-anesthetists (25 years)[a]						
P[b]	85	42	24	34	129	229
A	10	4	3	6	31	44
M	0	0	0	0	1	1 (28 years)
Female anesthetists (29 years)						
P	8	4	1	4	17	26
A	0	0	1	0	7	8
M	0	0	0	0	0	0 (33 years)
Wives of anesthetists (28 years)						
P	119	25[c]	21	17	74	137
A	11	5	5	4	14	28
M	1	0	1	0	2	3 (34 years)
All groups (28 years)						
P	212	70	46	54	215	392
A	21	9	9	10	52	80
M	1	0	1	0	3	4 (30 years)

Duration of occupational exposures at the time of the pregnancy

[a] The numbers in parentheses in first column indicate average age at conception for pre-exposure pregnancies and in last column for pregnancies during employment.

[b] P represents the total number of pregnancies; A represents the number of spontaneous abortions and perinatal deaths; M represents the number with malformations.

[c] One of these births was of twins.

"community" as all persons exposed to anesthetic gases in operating rooms in the U.S. and Great Britain who were members of their professional associations.

Dr. Knill-Jones of the University of Glasgow headed the study team in the U.K. and Dr. Ellis Cohen of Stanford Medical School headed the study organized by an ad hoc committee of the American Society of Anesthesiologists (ASA) in the U.S.

In the U.K. 1241 women anesthetists were identified and sent questionnaires.[3a] Their names were obtained from the 1970 records of the Department of Health and the Scottish Home and Health Department, the Association of Anesthetists of Great Britain and Ireland, and the Faculty of Anaesthetists of the Royal Colleges of Surgeons in England and Ireland. The experience of these women anesthetists was compared with a one in eight sample of women physicians in the Medical Register of the General Medical Council's branches in England, Wales, Scotland, and Ireland.

Exposed male anesthetists were subsequently studied.[3b] Questionnaires were sent to all Fellows in the Faculty of Anaesthetists, members of the Association of Anesthetists, and anaesthetists on the 1972 superannuation records of the Department of Health and Social Security and the Scottish Home and Health Department. For comparison, every tenth male doctor in the Medical Register and resident in the U.K. was sampled.

In the U.S. study[4] the questionnaire was sent not only to the members of the American Society of Anesthesiologists (male and female) but to the members of the American Association of Nurse Anesthetists (AANA), the Association of Operating Room Nurses (AORN), and the Association of Operating Room Technicians (AORT). These latter two groups are combined in tabulations (Table 3, AORN/T).

Table 2
SEX RATIO OF CHILDREN BORN TO DANISH COUPLES
ONE OF WHOM WAS EXPOSED TO ANESTHETIC GASES

Group	Pre-exposure	Duration of exposure at the time of pregnancy (years)				Total exposed
		1st	2nd	3rd	3+	
Nurse anesthetists						
Births	75	38	21	28	98	185
Males	45	19	8	13	53	93
Females	30	19	13	15	45	92
Female %	40.0	50.0	61.9	53.6	45.9	49.7
Female anesthetists						
Births	8	4	0	4	10	18
Males	5	0	0	2	3	5
Females	3	4	0	2	7	13
Female %	37.5	100.0	—	50.0	70.0	72.2
Wives of anesthetists						
Births	108	21	16	13	60	110
Males	55	7	9	8	25	49
Females	53	14	7	5	35	61
Female %	49.1	66.7	43.8	38.5	58.3	55.5
All exposed women						
Births	83	42	21	32	108	203
Males	50	19	8	15	56	98
Females	33	23	13	17	52	105
Female %	40.0	54.8	61.9	53.1	51.9	51.7
All exposed groups						
Births	191	63	37	45	168	313
Males	105	26	17	23	81	147
Females	86	37	20	22	87	166
Female %	45.0	58.7	54.1	48.9	51.8	53.1

For comparison purposes, questionnaires were also mailed to members of the American Academy of Pediatrics and a 10% sample of the membership of the American Nurses Association. A total of 73,496 individuals were sent questionnaires, of which 49,585 were presumably exposed and 23,911 presumably unexposed.

III. OBJECTIVES OF THE STUDY

The objectives of the study were (1) to determine if there were effects of trace anesthetic agents on health of operating room personnel and (2) to determine what proportion of these effects can be eliminated by reduction of exposures.

IV. METHODS

The work of the two groups was coordinated and similar methods used in each country. A mailed questionnaire was the prinicipal method for obtaining information. In the U.S. survey it was headed "Effects of Waste Anesthetics on Health". Questions were asked about involuntary infertility, about exposures of either parent, about abortion or other un-

Table 3
RESPONSE RATES FOR STUDIES OF HEALTH
EFFECTS OF ANESTHETIC GAS EXPOSURES IN THE
U.S. AND IN THE U.K.[3b,4]

Group	Number of question-naires mailed out	Number of usable replies	Percent of usable replies
U.K. total	7,949	5,549	70.1
U.S. total	73,496	40,044	54.5
ASA			
Men	9,793	6,558	67.0
Women	1,399	1,059	75.7
AAP			
Men	7,024	2,893	41.2
Women	886	639	72.1
AANA			
Men	2,627	1,894	65.4
Women	11,967	7,136	59.3
AORN/T			
Men	1,666	891	53.5
Women	22,133	12,272	55.4
ANA			
Men	320	142	44.3
Women	15,681	6,560	41.8

Note: The professional society abbreviations stand for: ASA, American Society of Anesthesiology; AAP, American Academy of Pediatrics: AANA, American Association of Nurse Anesthetists; AORN/T, Association of Operating Room Nurses and Association of Operating Room Technicians; and ANA, American Nursing Association.

favorable pregnancy outcome, the ages of the parents at the time of any pregnancies, smoking histories during pregnancy, the possibility of X-ray exposures or exposures to German measles (Rubella) during pregnancy, and whether either parents or children had cancer. Parents were asked questions concerning liver or kidney disease.

In most of the tabulations data have been "adjusted" so as to reflect groups who were comparable in terms of age at the time of pregnancy and in terms of smoking during the pregnancies.

The U.K. data on obstetric history were analyzed according to whether or not the respondent worked in the operating room (exposed) during the first 3 months of the pregnancy. No distinction was made as to whether the exposed person was a surgeon or an anesthetist (9% of the toal population were surgeons, 26% anesthesiologists, 39% did not practice in a hospital, 17% were other hospital staff members, and 8% had miscellaneous specialties).

In the U.S study, being exposed for females was defined as working in the operating room during the first 3 months of the pregnancy and during the previous calendar year. Exposures of males was defined on the basis of work in the operating room during the calendar year prior to the pregnancy. The replies of the various professional societies were tabulated separately. Social security numbers were used to avoid duplication of subjects.

V. RESULTS

Table 3 shows the response rates for the U.K. and U.S. populations.

In the U.K., study of women anesthetists compared to other women practitioners,[3a] there were no significant differences in sex ratio, stillbirths, or neonatal deaths. When the data

Table 4
FREQUENCY OF REPORTED CONGENITAL ABNORMALITY AMONG MALES AND FEMALES EXPOSED TO ANESTHETIC GASES IN THE U.K.[3b]

Parental exposure	Number of births	Abnormalities					
		Major	%	Minor	%	Total	%
Paternal	5175	56	1.08	160	3.09*	235	4.54*
Maternal	438	7	1.59	14	3.19	21	5.48*
Neither	6442	68	1.05	152	2.35	233	3.62

Note: * Signifies that the result is significantly different from the "neither exposed" with $p < 0.05$.

for anesthetists were analyzed according to whether the anesthetist worked during the first or second trimester of pregnancy, the number of reported congenital abnormalities was 6.5% of live births for the mothers who had worked, and significantly larger than the 2.5% for babies born to mothers who did not work during early months of pregnancy. The reported rate for other practitioners was 4.9%.

Working anesthetists had abortion ratios of 18.2% of pregnancies, significantly larger than the 13.7% for anesthetists who did not work during early pregnancy. For the group of practitioners, the ratio was 14.7%; 12% of anesthetists reported infertility with no known cause compared to 6% of the comparison group.

In the analysis of exposed males in the U.K.[3b] there was no apparent influence of paternal exposure on spontaneous abortions among their wives. However, as shown in Table 4, there did seem to be an effect on congenital abnormalities, primarily among minor ones, that is those which are not life threatening.

Results similar to those in Table 4 were obtained with data matched to reduce the influence of age and smoking differences which may have been present.

Exposure had no apparent effects on infant mortality or on the incidence of cancer or leukemia in the children.

Table 5 shows data on the reported frequency of abortion for the exposed and unexposed members of the U.S. professional groups.

When the congenital abnormality rates are calculated only for selected, relatively severe abnormalities (Table 6), and anesthetists are compared with pediatricians, the female respondents (exposures of females) report 1.24% for ASA members vs. 0.21% for APA members. The comparable data for male respondents (male exposure) are 1.56% and 0.90%. Because the sample sizes are larger for the latter difference, it has a lower probability of being due to chance (0.03) than does the former difference (0.06).

No data on unexplained infertility or on infant mortality sex ratios, or childhood cancer are given for the U.S. study.

Significant excess cancer was reported for exposed female respondents but not male respondents in association with exposure. For both sexes, exposure was associated with excess reported liver disease, even when serum hepatitis was excluded.

VI. INTERPRETATION OF THESE DATA

Both studies agree that exposure of women during the early part of pregnancy, and possibly during earlier periods increased risks for spontaneous abortion. There is a suggestion of a

Table 5
PERCENT OF REPORTED SPONTANEOUS ABORTIONS FOR PREGNANCIES AMONG POPULATIONS OF U.S. PROFESSIONAL SOCIETIES EXPOSED AND UNEXPOSED TO ANESTHETIC GASES ACCORDING TO WHICH PARENT WAS EXPOSED (SMOKING AND AGE STANDARDIZED)[4]

Organization	Mother exposed	Father exposed	Unexposed	
ASA	17.1 (468)	11.6 (3416)	15.7 (138)	
AANA	17.0 (1826)	11.7 (1350)	14.4 (676)	
AORN/T	19.5 (2781)	18.4 (237)	15.1 (1533)	
			Father	Mother
AAP	—	—	12.6 (1982)	8.9 (308)
ANA	—	—	10.0 (54)	15.1 (1948)

Note: Number of pregnancies given in parentheses. Professional society names as given in Table 3.

Table 6
PERCENT OF REPORTED CONGENITAL ABNORMALITIES AMONG LIVE BIRTHS TO POPULATIONS OF U.S. PROFESSIONAL SOCIETIES EXPOSED AND UNEXPOSED TO ANESTHETIC GASES ACCORDING TO WHICH PARENT WAS EXPOSED (SMOKING AND AGE STANDARDIZED)[4]

Organization	Mother exposed	Father exposed	Unexposed	
ASA	5.9 (384)	5.4 (2988)	3.4 (116)	
AANA	9.6 (1480)	8.2 (1168)	5.9 (566)	
AORN/T	7.7 (2210)	6.4 (203)	7.0 (1275)	
			Father	Mother
AAP			4.2 (1714)	3.0 (276)
ANA			3.7 (49)	7.6 (1629)

Note: Number of live births given in parentheses.

dose-response relationship in that the more heavily exposed anesthetists show more effects than do operating room nurses or technicians.

There is evidence of an effect on congenital abnormalities when either parent is exposed, but in the British study the evidence is more convincing for "minor" abnormalities, while the differences are close to conventional levels of significance for exposures of either parent for a group of serious abnormalities.

One difficulty in interpretation is the striking variability of the rates of abortion and congenital abnormality in unexposed members of the health care professions. Possibly there are other undetected risk factors than smoking and age, for which adjustments were made.

The findings are sufficient to indicate three things: *women should not be exposed to work involving exposures to waste anesthetic gases during early pregnancy; waste anesthetic gases can and should be controlled in order to protect the health of all operating room personnel; prospective monitoring of reproductive outcomes of nurses and of operating room personnel is needed to assure that unfavorable effects do not continue to occur.*

VII. SUBSEQUENT STUDIES FOR WHICH THE NEED WAS INDICATED BY THIS RESEARCH

Three types of studies have been done to shed additional light on the findings from this community study of reproductive outcomes of pregnancies in which one parent was exposed to anesthetic gases. These are methodological studies of the bias which may occur when such data are obtained by mail questionnaire. A second type of study is supplemental laboratory tests on experimental animals to see what the effects of their exposures may be on reproductive outcomes. Finally, additional epidemiological studies of other exposed populations are indicated.

Axelsson and Rylander[5] used a postal questionnaire among women who had worked in a Swedish hospital, and then examined the data to see if there were any excess numbers of abortions among women exposed to anesthetic gases during the first trimester of pregnancy. They found a higher, but not significantly increased rate among exposed women. They then examined the hospital records for all the women surveyed. All of the women who had a miscarriage and were exposed during the first trimester reported both work exposure and miscarriage, but a third of all miscarriages occurring to women who were not exposed were not reported on the questionnaire. When the 113 pregnancies from nonrespondents were added to the 655 pregnancies reported as a result of two rounds of mailed questionnaires, the miscarriage rate in the exposed was 15.1% compared to 11.0% among the unexposed, a difference which was not considered to be significant. Discrepancies were also found with respect to week of pregnancy during which miscarriage occurred.

One single study of attempted laboratory confirmation will be cited.[6] It was a study of male mice exposed to either 0.5%, 5%, or 50% nitrous oxide for 14 weeks, with air inhalation as a negative control and exposure to methyl methanesulfonate, a known spermatotoxin, as a positive control. There were no significant differences among the exposure group and the negative control group with respect to testicular weight, sperm counts, proportion of abnormal sperm, or histological appearance of the testes.

Sperm morphology and sperm counts have been compared in anesthetists working in operating rooms with good control of waste gases, and no abnormalities have been found.[7]

Finally, another large "community" study has enquired of dentists and dental assistants, some of whom use anesthetic gases and some of whom do not, as to their reproductive histories.[8] The same team that studied operating room exposures in the U.S. carried out the study. The response rates for the study were 73.6% for the dentists and 70.0% for the assistants.

As is shown in Table 7, both wives of male dentists and female assistants had increased rates of spontaneous abortion, and the rates were greater for those who reported heavy use of anesthetic agents than for light users. By contrast, as is shown in Table 8, no increase in congenital abnormality rates is found for children of the male dentists, but significant increases are found for light but not for heavy users among the children of assistants.

Typical values of commonly used anesthetic gases in operating rooms are halothane in the range of 10 ppm and nitrous oxide at levels about 600 ppm. Scavenging systems for waste gases can reduce these levels to 0.05 ppm for halothane and about 1 ppm for nitrous oxide. These figures apply to "closed systems", which are usually used in operating rooms. However the technical procedures for open systems, as are usually used in dental anesthesia are somewhat more difficult. The predominant exposure in dentistry is to nitrous oxide (N_2O). When the data are analyzed by type of gas used, for both dentists and chairside

Table 8
PERCENT OF CONGENITAL ABNORMALITIES REPORTED BY DENTISTS AND ASSISTANTS ACCORDING TO WHETHER THEY WERE NONUSERS OF ANESTHETIC GASES, LIGHT USERS, OR HEAVY USERS (AGE AND SMOKING ADJUSTED)[8]

Use category	Wives of dentists (paternal)	Female assistants (maternal)
Nonusers	4.9 (5277)	3.6 (2882)
Light users	4.6 (1890)	5.7 (341)[a]
Heavy users	4.8 (1177)	5.2 (316)

[a] These values are all significantly different from those of nonusers with p <0.05. Numbers of live births in parentheses.

Table 7
PERCENT OF SPONTANEOUS ABORTIONS ACCORDING TO PATERNAL OR MATERNAL EXPOSURES TO LIGHT AND HEAVY USE ANESTHETIC GASES IN DENTISTRY (ADJUSTED FOR AGE AND SMOKING)[8]

Use category	Wives of dentists (paternal)	Female assistants (maternal)
Nonusers	6.7 (5709)	8.1 (3184)
Light users	7.7 (2104)[a]	14.2 (407)[a]
Heavy users	10.2 (1328)[a]	19.1 (400)[a]

[a] These values are all significantly different from those of nonusers with p <0.05. Numbers of pregnancies in parentheses.

assistants it is nitrous oxide alone which is associated with not only unfavorable reproductive outcomes but most likely to neurological disease as well.

The Ad Hoc Committee on the Effect of Trace Anesthetics on the Health of Operating Room Personnel of The American Society of Anesthesiologists urged that a follow-up study be done to assure that the effects which were documented were abated as controls were introduced to collect waste gases in surgical operating rooms. The Council on Dental Materials, Instruments and Equipment of the American Dental Association recommended that effective scavenging devices be installed in dental offices using inhalational anesthesia, and that a monitoring program be instituted in offices where nitrous oxide is in use.

Have these recommendations been followed? Ask your dentist, surgeon, or anesthetist.

Although the risks for health damage of the community at large are small, the awareness of the risks to themselves and their families may add caution to the practices of dentists, surgeons, and anesthetists.

REFERENCES

1. **Vaisman, A. I.,** Working conditions in surgery and their effect on the health of anesthesiologists (in Russian), *Eksp. Khir. Anesteziol.*, 3, 44, 1967.
2. **Askrog, V. and Harvald, B.,** Teratogen effekt af inhalationsanaestetika (in Danish), *Nod. Med.*, 16, 498, 1970.
3a. **Knill-Jones, R. P., Moir, D. B., Rodrigues, L. V.,** et al., Anaesthetic practice and pregnancy: a controlled survey of women anesthetists in the United Kingdom, *Lancet,* 2, 1326, 1972.
3b. **Knill-Jones, R. P., Newman, B. J., and Spence, A. A.,** Anaesthetic practice and pregnancy: controlled survey of male anaesthetists in the United Kingdom, *Lancet,* 2, 807, 1975.
4. **Cohen, E. N.,** (Chairman), Ad Hoc Committee on the Effect of Trace Anesthetics on the Health of Operating Room Personnel, American Society of Anesthiologists, Occupational disease among operating room personnel, *Anesthesiology,* 41, 321, 1974.
5. **Axelsson, G. and Rylander, R.,** Exposure to anesthetic gases and spontaneous abortion; response bias in a postal questionnaire study, *Int. J. Epidemiol.*, 11, 250, 1982.
6. **Mazze, R. I., Rice, S. A., Wyrobek, A. J., Felton, J. S., Brodsky, J. B., and Baden, J. M.,** Germ cell studies in mice after prolonged exposure to nitrous oxide, *Toxicol. Appl. Pharmacol.*, 67, 370, 1983.
7. **Wyrobek, A. J., Brodsky, J. B., Gordon, L., Moore, D. H., Watchmaker, G., and Cohen, E. N.,** Sperm studies in anesthiologists, *Anesthesiology,* 55, 527, 1981.
8. **Cohen, E. N., Brown, B. W., Wu, M. L., Whitcher, C. E., Brodsky, J. B., Gift, H. C., Greenfield, W., Jones, T. W., and Driscoll, E. J.,** Occupational disease in dentistry and chronic exposure to trace anesthetic gases, *J. Am. Dental Assoc.*, 101, 21, 1980.

Chapter 21

OCCUPATIONAL HEALTH AS A COMMUNITY PROBLEM

John R. Goldsmith

While it may not be customary to link occupational and community health problems, it can be shown that there are at least three different ways in which occupational health is related to community environmental health. Occupational exposures are usually the best defined and most intense exposures for which human experience is available and therefore the data obtained or obtainable from occupational epidemiological studies have a unique place in the effort to evaluate community environmental health problems.[1] So the first linkage is the linkage related to specific hazardous materials. It must be emphasized that the occupational (working) population does not represent the general population in that it contains people in good enough health to hold regular jobs. In the general community there are ill persons, elderly persons, as well as infants and children who may be far more seriously affected by a pollutant exposure which would hardly affect a healthy workman. Nevertheless, there is hardly a pollutant known, for which qualitative and semiquantitative information about health risks is not derived from the experience of a group of persons exposed as a consequence of their occupation.

Secondly, agents which may impair the health of workmen also occur in the general community either because the general community is exposed when using the materials or when exposed to industrial waste gases, solids dust, or liquids. Thus, DBCP production workers are not the only group among which male fertility impairment has been detected. Those who apply the material may be affected as well, and since it tends to persist in ground water, we shall have to face the fact that the public in certain areas may be exposed to DBCP, a known carcinogen and mutagen, in drinking water. Anesthetists are not the only persons exposed to anesthetic agents; their patients are as well, even though they are not likely to be exposed to the same frequent low-level exposures as are anesthetists and other operating room personnel.

Asbestos workers are not the only community members at risk from long-term effects of asbestos. Their wives, or others who laundered their clothes are at risk as well. Asbestos waste, which used to be handled carelessly when less was known as to its long-term health risks, may contaminate play areas, walk ways, and gardens, leading to substantial exposures to community members who are unaware of the risks.

The most widely recommended method for protecting the health of workmen is "adequate exhaust ventilation" of the workplace. But exhaust ventilation is a practice of transferring the hazardous material from inside the plant or factory to outside air, where the possibility of community exposure must be (and usually is not) considered. Occasionally an error or explosion in the plant or distribution facilities occurs and community exposures may be widespread with a wide variety of complex reactions such as the Kepone disaster in the James River Estuary of Virginia,[2] the Seveso incident in Italy,[3] and the contamination of dairy feed by PCB (polybrominated biphenyls) in Michigan and the Bhopal tragedy in India.[4] In general we have to assume that whatever may affect the health of workmen may also affect the health of the community as well. Merely because it is more likely to be diluted by the time most community exposures occur, we cannot assume that it will have no health effects in the community, because in general there are groups in the community who are unusually susceptible because of age and medical status.

Finally, occupational groups constitute a true community in the sense that they reflect a population with common problems, risks, and often actually live in the same neighborhoods.

The nature of the occupational community underlies the rationale for occupational epidemiology.

In fact, one of the weaknesses of conventional occupational health is the failure to recognize the community nature of occupational groups. If a peculiar form of dermatitis or headache on first coming to work on Monday affects one worker exposed to a given agent, the chance that other workers similarly exposed is good and they should be questioned or examined in order to see if there is not a widespread effect.

So similar are community and occupational environmental health problems with respect to agents, methods for detection and monitoring, and types of effects that it really makes no sense to separate training and research efforts in the two fields. The only reasons that such separations persist is that governmental programs, responding as they do to different constituencies at different times, have kept these two types of activities separate administratively, when logically they should be a single unified program, sharing personnel, laboratory and monitoring equipment, and epidemiological experience.

REFERENCES

1. **Monson, R. R.** *Occupational Epidemiology,* CRC Press, Boca Raton, Fla., 1980.
2. **Epstein, S. S.,** Kepone-hazard evaluation, *Sci. Total Environ.,* 9, 1, 1978.
3. **Pocchiari, F., Silano, V., and Zampieri, A.,** Human Health Effects from Accidental Release of Tetra-chlorodibenzo-p-dioxin (TCDD) at Seveso, Italy, *Ann. N.Y. Acad. Sci.,* 320, 311, 1979.
4. **Landrigan, P. J., Wilcox, K. R., Jr., Silva, J., Jr., Humphrey, H. E. B., Kauffman, C., and Heath, C. W., Jr.,** Cohort study of Michigan residents exposed to polybrominated biphenyls: epidemiologic and immunologic findings, *Ann. N.Y. Acad. Sci.,* 320, 284, 1979.

Section III: Environmental Epidemiologic Studies as a Basis for Health Protection

Chapter 22

ENVIRONMENTAL HEALTH DECISIONS: THE IMPACT OF EVIDENCE

John R. Goldsmith

TABLE OF CONTENTS

I. INTRODUCTION

The bottom line for a community study is not what has been learned but what action has been taken. We assume that facts can inform decisions which result in action.

Now that a carefully selected set of community studies has been presented, and for each, the implications have been spelled out if possible, it is time to examine the decision process and to ask how the facts which these community studies produced influenced them.

Environmental health decisions reflect political, historical, and economic considerations as well as what is known and/or suspected about the impact of environmental exposures on health. To undertake to conduct community studies without understanding some of these other factors is sure to be disappointing or frustrating or both.

Environmental health problems do not occur as a result of intention, but usually as a result of inattention. Nobody wanted to spread cholera in 19th Century London. Nobody wanted elderly men and women to die in heat waves in 20th Century Hollywood. Farmers and even chemical companies don't want to make chemical workers sterile, nor do Michigan tavern owners want to see teen-agers die on or off the roads.

Even the inattention which allows some of these tragic things to occur is not necessarily deliberate, callous, or indifferent, nor is it inherently negligent. The inattention is usually due to a failure to see and respond to what is often a new situation.

Photochemical smog in Los Angeles was a new provocative cause of asthma attacks; so were sulfur oxides and smoke in post World War II Kawasaki a new cause of respiratory impairment in children.

The effects of lead on the learning abilities of school children may have been going on for decades at least in urban cities, but our ability to detect these effects is new.

The uses of chemicals for anesthesia have been known from the middle of the 19th Century, in fact Dr. John Snow was one of the first anesthetists in England. (Fortunately for epidemiology he often had part of the day off for studying cholera.) Agents like halothane are new, as are long and intense exposures of numbers of young women — nurse anesthetists.

The new circumstances and situations, the newer methods for investigation and detection are likely to uncover a need for changing, adapting, or terminating a well established set of procedures. The need is there because under these new circumstances, continuing in the old established ways will allow environmental health problems to continue to grow. These new situations which require alteration in established ways of doing things create a political tension between established institutions and those concerned with newly recognized health problems.

Snow's opportunities to obtain and analyze data and to have the impact he had was partly based on Chadwick's report authorized by the British Parliament. Dr. William Farr's General Registrar's Office was a governmental function, and the political processes of mid-19th Century provided these auxiliary resources.

Some of Snow's recommendations were directed to persons at risk, some to health care providers, and some to governments. All of them accepted to a greater or lesser extent that changes were needed and there existed the mechanism to make them. Similarly, in Jerusalem, it was a governmental agency with the power to restrict the distribution of food supplies in the market which had the power to terminate the practice of generations of farmers of using human excreta for fertilizer of human food crops. It was also a governmental agency which had the authority to sample and test water and food. Since the provision of a good quality water supply to parts of Israel previously arid or dependent on wells was an important objective of the government, the laboratories and water monitoring programs were in place.

In Delano, when (state) governmental regulations required that mothers with new babies be warned that the public water supply was not good for them to use in making up bottles for their infants, they naturally enough looked to the government (city) for satisfactory water for the babies' formulas. The city government, not wanting to take on an unnecessary expense then turned to its legislator, who in turn asked the (state) health department to evaluate the problem. The fact that in Jerusalem it was a Jewish government and Arab farmers whose practices needed to change, and that in Delano it was an "Anglo" government which needed to modify its water supply policies in order to protect Hispanic babies makes no difference in the political nature of the change process.

The political decision processes of the 20th Century have involved the largest of industries. Agriculture (Chapters 3 and 19), motor vehicles (Chapters 10, 14, 15, 16, and 17), electrical utilities (Chapters 6 and 10), housing (Chapters 2, 6, 10, and 15), health care (Chapter 20), pulp and paper (Chapter 11), petrochemicals (Chapters 5, 8, and 19), and licensed beverages (Chapter 17) are among the involved industries. Internally, the components of such industries change in order to introduce new products, compete for a larger share of the market, or to increase profits. Being competitive as nearly all of them are, changes for the purpose of protecting the health of a community or of their workers is seen as an obstacle to either competition or profitability, therefore such changes must usually be imposed or required by outside authority, if they are to occur. Only a political process is capable of such impositions and requirements.

A historical understanding of the need for technological changes in order to protect health from environmental damage is more difficult to achieve than is a political understanding. As was said earlier, the people who run factories, utilities, and farms do not intend to cause health damage to the communities they affect. They are merely trying to do their productive work the best way they know. In many cases what has happened is that the scale of their activity got so much larger that it exceeded previous experience, or they were dealing with new and unprecedented materials or processes, so that their established ways of dealing with things were no longer appropriate.

A third historical process is that public awareness has increased, tolerance of pollution has decreased, and methods for detection and measurement have improved so that things which would have been overlooked or been undetected in previous eras are now identified as harmful.

The economic dimensions of these issues hardly need to be stressed. The industries affected are immense enterprises, with complex personal, political, and financial interrelationships. Space only permits a few vignettes.

Lead is a major nonferrous metal, but it is often found and mined from ore bodies including copper and zinc, so its price and use are linked with the prices and use of other nonferrous metals. (In some places precious metals are also part of the complex.)

Dr. Robert Kehoe as a young man was largely responsible for the industrial hygiene provisions which allowed for the safe manufacture, blending, and use of lead tetraethyl, the major source of lead as an air pollutant 20 to 30 years later. After these decades, Dr. Kehoe became the dean among occupational health physicians in the U.S. and the Medical Director of the Ethyl Corporation, which manufactured tetraethyl lead. It was very difficult for Dr. Kehoe to come to grips with evidence of a community risk from the widespread use of a product he had been so involved with. One of the economic complications was that lower grade fuels would provide satisfactory performance in automobile engines if sufficient tetraethyl lead were added. Thus, continued use of lead additives had an important economic impact not only on the Ethyl Corporation, but on the petroleum industry as a whole.

Once the possibility of a risk to community health of lead exposures became a matter for scientific evaluations, the medical directors of the major petroleum refiners let it be known that reducing or removing lead from motor fuel would not be opposed by them (or by inference by their companies). The reason was simple; the large refiners had huge catalytic cracking installations, a wide choice of petroleum stocks, and to upgrade the octane rating (and price) of their product was feasible. Smaller local refiners, many of whom were offering gasoline at lower prices already, did not have this flexibility. Thus, restrictions on lead in motor fuel would give a competitive advantage to the large companies. These same companies carried on extensive advertising campaigns, and many members of the public were convinced that they should use the best (and most expensive) gasoline they could buy for the cars on which they depended for so much. At one time, the California Health Department and California Air Resources Board surveyed the octane requirements specified by automobile manufacturers and the pattern of purchase of gasoline by octane level. They computed that the public was paying at least $50,000,000 more a year in California alone for unnecessarily high octane and high lead content gasoline which their cars did not need according to the manufacturers. The competition between large and small refiners has continued to be a major feature of the political (regulatory) decision processes affecting lead levels in gasoline.

New industrial facilities were the source of environmental hazards in Poza Rica and in Eureka. Possibly the same was true for Kawasaki. Freeways, in 1948, a new type of mass transit, were responsible for the emergence of photochemical smog in Los Angeles, since motor vehicle emissions were responsible for the bulk of the photochemical pollution and for the elevated carbon monoxide levels as well. These developments represented huge capital investments in their respective communities. The mere implication that they may have been responsible for community health hazards generated strong feelings. On the more positive side, the availability of air conditioners in nursing homes provided substantial protection of their elderly and frail residents from heat waves, about which despite the great numbers of excess deaths, no health authorities have yet taken with a suitable level of seriousness. Such air conditioners represent not only a large investment for the households and nursing homes but also for the electrical utility which provides them with power.

II. WHERE AND BY WHOM ARE THE DECISIONS MADE?

In one of the examples in the previous section, the voters of the state of Michigan decided that the minimum age for legal purchase of alcohol should be 21 years. Plant management in Eureka seems to have made decisions to reduce emissions (although that is a bit speculative). DBCP production was suspended by all producers without waiting for governmental action, once the evidence of male sterility in production workers was clear.

In a sense no decisions were made for reducing the mortality from heat waves or reducing the risk from anesthetic agents. Urban authorities in London and Jerusalem have taken steps to reduce the risks from cholera. The bulk of the remaining decisions were made by the national government or by state governmental agencies authorized to do so by the central government. Political pressure was the major mechanism urging action, but facts which were generated by the community studies led to this pressure.

III. ALTERNATIVES TO POLICE POWERS; LAW, MARKET FACTORS, AND MANNERS

In the U.S., it is considered conventional wisdom to look to enforceable regulations as a basis for environmental health protection. The police powers of government are considered the preferred mechanism by which communities can improve and sustain environmental quality. Jail and fines and their threats are considered the mechanisms of choice.

There are other ways, as they say, to skin a cat. "Effluent charges" represent an effort to reflect the cost for emitting pollutants as a cost of doing business. It would operate in such a way that a firm emitting pollutants is charged by the community for its use of the scarce diluent properties of the air or water to which the emissions are made. Under the usual circumstances, those of us who have to breathe the air or drink the water risk our health because a pollutor has gotten to use the diluting properties of the air or underground water before we had a chance. Those risks are what economists term "externalities" for the firm which is emitting the pollution, because the costs are borne by the public, or are in event by an entity that is external to the firm. The use of an effluent charge is a way to "internalize" the costs and provide an economic incentive to the firm to keep its pollution down. For further information, see *Air Conservation,* a publication of the American Association for the Advancement of Science.[1]

The most favorable type of influence is the influence of public opinion or public acceptability. It is perhaps too much to expect that soon firms and individuals will recognize that each of us has a right to good water, a safe workplace, and clean air, and that it is irresponsible to infringe that right. In a sense, the ultimate guidelines are those best designated as manners. In a world of harsh economic competition, good corporate manners may sound like more than one can expect, but as we said earlier, *no technical process manager wants to poison the neighbors.* Many in fact would prefer that their operation was considered to be beneficial to the public and to the community in which their employees and customers live. They want to behave well and be liked by their neighbors. Community organization can make a strong input to the realization of that possibility. The main benefits of governmental regulations are to define what is considered to be acceptable by the community, and to reduce the price or economic advantage which firms can obtain by externalizing (that is by passing on to the community and environment) the costs and risks of the pollutants they produce.

Thus, if governmental regulation is to be considered the fallback strategy for maintaining a satisfactory level of environmental quality, at the very least let us emphasize that it may have at least two mechanisms; (1) to punish offenders and threaten potential offenders and (2) to provide guidance for the managers who really want to be good neighbors, who care about the wholesomeness of the communities where they operate and live. Adversary relationships and threats may be needed, but the mobilization of good will and good intentions are the ultimate objectives for maintaining a satisfactory level of environmental quality.

IV. THE COMMUNICATION PROCESS; DISTORTION AND THE MEDIA

Pollution, new disease risks, and the impact of technology are newsworthy. But news headlines do not always provide the optimal input to a complex community decision. The community needs a way of informing itself, and in most communities, newspapers, radio, and television are the mechanisms for meeting these needs. Most reporters feel that extra deaths are more interesting than additional cases of illness, and they can seem very bored to hear about objections due to odor. Thus, the influence of the "media" can tend to distort the proper balance of judgment as to which types of reactions may be most important.

A relatively recent development is the professionally qualified science writer. Such reporters can be an immense asset in communications to other members of the community and to the governmental agencies. Science writers have been invaluable in transmitting to the public a balanced picture of the risks of pollution and what is known as well as what is obscure about the sources and mechanisms. They can also help the community health worker to better understand the feelings and concerns of the community.

There may be a risk of oversimplification or distortion of environmental problems, but usually it is a risk worth taking in the interest of better informing the community as to the problems. A well informed community is more likely to be a community whose political wishes are listened to in the congress, legislature, or council chamber. In short, those wishing to act effectively against community pollution problems have in the "media" a potentially powerful ally, especially if the reporters are qualified science writers.

V. INERTIA IS THE OBSTACLE; FACTS ARE THE TOOLS

If one rejects the "devil theory" of environmental pollution, then how and where do we focus our efforts, our energies, and to whom do we direct our facts.

In this complex world, as in the simple one we are so nostalgic for, most of our actions are habitual, not innovative. We eat, we dress, we speak "How do you do?", we smoke, we drink, we drive, we work, and do all of these things the way we do because of habit. To change one of these habits is very difficult. For a community, for example to change its habitual way of disposing of leaves, by burning them, is difficult. There is a great deal of inertia in everything that we do, and if some of these actions produce a pollution problem in our community, it is difficult to change that too. We have elaborate ways of avoiding change. Let us consider some of the foolish games we play rather than decide to act against pollution in ways which can be effective.

The first kind of game is "*you* are polluting my environment". This is the devil theory in a new guise, and it's escapism. All of us are involved to some extent in the community decision process, and none of us possess the environment as our personal property. Sure we can find someone else to blame, but that is a game we play to avoid being responsible community members. Of course there are pollutors and they may need to be put under pressure to accept their proper responsibilities, but it is partly up to others in the community to help define that responsibility.

A second type of game is played around the theme "It's new, it's bigger, and therefore it must be better". This game is built around gullibility to the huckster. It leads to rapid and early obsolescence and the need to dispose of huge amounts of waste. Of course we like to live well and enjoy a good living standard, but we must also insist that this is not at the expense of environmental degradation which may handicap the health of our children and our grandchildren. We need not settle for dirty old clothes and dilapidated houses, but we do need to examine the care with which the material through-put that is the basis for our living standard is managed.

Another game which pollutors and health authorities are fond of playing is that "pollution is your privilege until it can be *proven* that health is harmed". The most cost-effective time to prevent pollution is in the planning stage, and this game lets this valuable time pass. Of course expenditure of money to prevent pollution deserves to be justified, but it may take decades to "prove" that cancer or chronic respiratory disease is caused by air pollution or that impairment of learning capabilities is caused by elevated levels of methemoglobin in the blood or lead in the play area. Risking a generation of children with intellectual handicaps is a frightening prospect and may prove to be very expensive in terms of lost productivity, and the care of the mentally or emotionally subnormal. Would it not be preferable to reverse the "proof" game and ask pollutors to prove that their emissions are so low that there is no risk that they will harm health? The strategy of epidemiological monitoring is one such approach.

Some industrial managers play games when they say: "We are complying with all the laws and giving the public a superior product to meet their demand." Often that demand is manipulated and only the industry is assumed to be capable to knowing what is being demanded. In fact, in survey after survey, the public has shown its willingness to pay the reasonable costs of prevention of pollution. Some industries comply with the laws as slowly and with as little initiative as their legal departments feel they can without the executives being fined or imprisoned. Many industries have recognized that long court fights against regulations allow them to defer capital expenditures and thus at the minimum save on the interest which otherwise they could get (or have to pay).

Another game all too frequently played by representatives of industry is "these tough regulations will force us to close our plant and throw thousands out of work". This type of economic blackmail should not be tolerated. If it is profitable for the company to make a product, in the long run it is profitable to make it in a way that is consistent with the public's health and well-being. If you get a company manager in a relaxed mood you often learn that the plant in question is obsolete and no longer profitable, or that the threat was an empty one. Community studies of environmental health problems are a way to generate facts about health and environmental risks and put them on record. Few plant managers who make threats of closure due to costs of environmentally indicated regulations are willing to document their claims that the regulations will "put the company out of business" on the open record. They must be made to realize that unless they are willing to do so, the threat won't work.

A number of governments play a foolish game of "We have passed the toughest standards of any" by which they mean that they have chosen a lower number than other agencies. It rarely means that the low number commits the governmental agency to an adequate program of enforcement or prevention to meet the low standard. What counts is keeping the pollution from being an environmental threat, not choosing a low number.

Another game that governments are fond of playing is to establish monitoring for pollutants instead of preventing their emissions. Often the location and calibration of the monitoring apparatus are not such that the numbers they generate mean anything concerning the risk to health or the environment, but a gullible public is given the impression that something is being done, when really nothing is being done. In the worst situation, the monitoring data are not analyzed or reacted to in any effective manner, and the public is given a sense of false security.

Another of the favorite games to keep us from having to change our polluting ways is the game of "Speak loudly, but don't drink the water and don't breathe the air!". Alternatively, this is a way of saying let's make a big commotion, but let the public beware; let the mothers buy bottled water and the scientists of 20 years hence can find out if too much nitrate in the water increases the cancer risk. We need to understand that our actions and our complaints need to have a balance.

Finally, we are all tempted to play the game of "These problems are too difficult and effective action would be too disruptive to our present way of life". The effect of this evasion is to pass on to our children the polluted environment and the responsibility of doing something effective about it. In the meantime if the smog level in Los Angeles gets too high, we keep the children from playing out-of-doors! This doesn't seem like a very responsible course of action — to restrain our children when we haven't the capacity or will power to restrain our own uses of automobiles so that their emissions will not threaten the health of our children.

There is only one game worth playing — the game of life. We must understand the impact of our own behavior, we must overcome our inertia so as to modify this behavior to the end that our fellowmen and our children will have a more wholesome world to live in and to enjoy.

The studies described in the preceding section have generated a uniquely valuable and useful set of facts. In most of these communities these facts were sufficient to overcome the inertia which allowed pollution to get out of hand. Both the methods and the facts make it easier for the next community facing similar problems to overcome its inertia.

REFERENCE

1. American Association for the Advancement of Science, *Air Conservation,* AAAS, Washington, D.C., 1965.

Chapter 23

THE ECOLOGICAL IMPERATIVE: SCIENCE AS A PROTECTIVE SYSTEM AGAINST UNINTENDED EFFECTS OF TECHNOLOGICAL CHANGE

John R. Goldsmith

Human health is not the only hostage to technology. Birds and grass, rats and mice, falcons and pelicans, lichens and orchids, lettuce and cotton were also threatened. We did not set out to write a book about the ecological impact of pollution, but no book on the human impact can be complete without recognizing the scale and complexity of the effects of pollution on the natural world which sustains human life and for which we must accept responsibility.

The scientific disciplines involved are numerous — meteorology, oceanography, limnology, biology and biochemistry, genetics, botany, horticulture, — and the list can easily be doubled or tripled. All of this scientific activity and work is needed in order to help us to understand the consequences of continued expansion of the pox of pollution, and to take meaningful action. Experimental research forms one sector of this scientific enterprise, and observational research a second sector. It is important to understand that if we want to try to determine whether the increase in fossil fuel combustion that releases megatons of carbon dioxide to the atmosphere results in an increase in the earth's temperature, we can hardly do an experiment. We must undertake observational research, and develop and refine models based on our observations. From these models we can attempt to make predictions and hopefully to test them. A related problem is the possible effect of increased particulate emissions on the reflectivity of the atmosphere, which may also have an effect on the heat balance of the earth, but in the opposite direction. Volcanic eruptions constitute a natural source of large amounts of particulate matter, so from observational research on the effects of volcanism, we may be able to draw reasonable inferences and make testable predictions as to the possible effects on climate of increased particulate emissions.

One of the consequences of such modeling was the prediction that large amounts of the relatively inert fluorohydrocarbons (Freon®) formerly used as propellants in spray can packages, could react with stratospheric ozone to reduce its concentration and thus allow increased penetration to the surface of the earth of energy-rich UV light. Such a prediction was sufficiently alarming that the use of such propellants has been sharply reduced, at first voluntarily and then as a result of regulation. For issues like these, all of the people on earth are as one community, and the only effective regulations must have international recognition regardless of how they are enforced.

On a somewhat more modest scale, the deaths of horses in the vicinity of the town of Benicia situated near the entrance of the Sacramento River into San Francisco Bay, had repercussions which illustrate the ecological consequences of pollution. In January 1970, a rancher who has pastured horses for many years reported that a number of horses had died. Since a new and very large petroleum refinery had recently been put into operation within a few miles, he wondered if the refinery could have affected his horses. When the horses were first examined the report was that they had died of anemia, but the rancher was not satisfied since before they died they seemed to have trouble breathing, and anemia shouldn't have had that kind of an effect. When the California Health Department staff visited the area of rolling hills, they could see a large number of possible sources of pollution: oil refineries, a place where old railroad boxcars were burned to salvage the metal, a dump where sludge from the manufacture of tetraethyl lead (among other things) was deposited, and a few miles away the slender tall stack of the nonferrous metal smelter at Selby. When samples of the blood and tissues of the horses were brought into the health department's laboratory they were found to contain very large amounts of lead. This finding immediately aroused the suspicion that the smelter emissions might have caused lead poisoning in the horses; the diagnosis was entirely consistent with the manner in which the horses died, since lead seems to affect the nerves of the larynx in the horses, causing them to have a croupy type of difficult breathing.

A plant pathologist was the next scientist to contribute to the situation. She found that the head of the wild oat plant *(Avena fatua)* after it has shed its seeds is an excellent indicator of airborne lead and other very fine particulate matter. The heads of the oats from the pastures where the horses had died had very high lead levels. We then faced two other problems; where did the lead come from and could it be taken up by the roots or was it primarily airborne. With the help of a mathematically oriented meteorologist, we then identified the six possible causes of lead pollution in the pasture, and the expected pattern of occurrence of lead levels in *Avena* which might indicate the most likely source of lead. The six possible sources were the oil refinery, emissions from motor vehicles crossing the nearby highway bridge, lead paint from the bridge, which from time to time had to be sand-blasted and replaced, the lead paint from the old boxcars which were burned, the tetraethyl sludge dump, and the smelter. The pattern of distribution was only consistent with one of the sources, the smelter. A similar pattern of distribution of cadmium in the *Avena* samples, which could only have come from the smelter clinched the matter. We looked at lead levels in other grazing animals, but they were not elevated. This turned out to be due to the tendency of horses to graze grass down to the roots, which other grazing animals did not do.

Then plots of *Avena* were grown in soil from the pastures where the horses had died, and other plots with soil taken from a remote rural location were grown on the pasture. *Avena* grown on the soil from the pasture did not have high lead, but plants grown on fresh soil but exposed to the winds carrying the smelter emissions did have high values. This satisfied us the lead was an air contaminant, rather than a soil contaminant. Cadmium, by contrast was readily taken up from the soil.

These findings alarmed the health and air pollution authorities who then authorized the study of children and adults in the vicinity in order to see if they had any evidence of health risk from lead.

The Selby smelter dated back to the gold rush days, and in 1915, horse deaths from lead poisoning had occurred in the same vicinity and a rancher had sought and obtained a permanent injunction against the emission of lead fumes from the smelter!

Although lead is apparently not harmful to the oat plant, it makes it toxic for one grazing species but not for others. We also investigated the distribution of small rodents in the vicinity and found that the numbers were lower than expected in the most heavily lead polluted areas. The livers of animals which were caught in the polluted pasture had high lead levels and their kidneys had high cadmium levels.

In the autumn of the year when the children returned to school we began a study of their lead levels and looked for biochemical evidence that there was a harmful effect from the levels to which thay had been exposed. Just as we were starting, the Historic Old Selby Smelter was shut down, because according to the plant management, the smelter was obsolete. Despite the fact that the smelter had been shut for several months before we took our samples, the children had relatively high blood lead levels compared to levels among other school children with similar contemporary exposures to lead in air. These results are shown[1] in Table 1 and Figure 1.

The lesson that this experience teaches is the value of the ecological perspective in protecting human health.

Without the veterinary pathology and plant pathology, the risks of increased body burden of lead in local school children would not have been identified. Without the meteorological

Table 1
AMBIENT AIR LEAD LEVELS AND BLOOD LEAD LEVELS IN THIRD GRADE CHILDREN IN CALIFORNIA, BY SEX FOR TWO NORTHERN CALIFORNIA COMMUNITIES NEAR A RECENTLY SHUT-DOWN SMELTER, AND FOR SOUTHERN CALIFORNIA COMMUNITIES WITH VARYING EXPOSURE TO MOTOR VEHICULAR POLLUTION

| | | Blood lead in μg/dℓ | | | | |
| | | Boys | | Girls | | |
Community	Mean air lead (μg/m³)	No.	Lead	No.	Lead	Mean lead
Smelter						
Benicia	0.29[a]	17	13.1	17	13.7	13.4
Crockett	0.28	18	14.3	10	13.8	14.1
Other						
Burbank	3.36	17	23.3	19	20.4	21.8
Riverside	2.75	27	12.1	23	11.4	11.8
Manhattan Beach	2.41	21	16.8	19	17.1	16.9
Azusa	2.00	18	12.6	23	12.0	12.3
Culver City	1.93	28	14.6	21	12.4	13.7
Long Beach	1.69	22	11.4	24	10.7	11.0
Oceanside	1.33	24	10.8	19	9.8	10.4
Lancaster	0.77	10	9.2	28	8.5	8.7

[a] Ambient air levels are mean values for available high volume samples. The data for Benicia and Crockett are measured from samples taken after the smelter had shut down. For locations of the communities in the "other" group, see Figure 1 of Chapter 9.

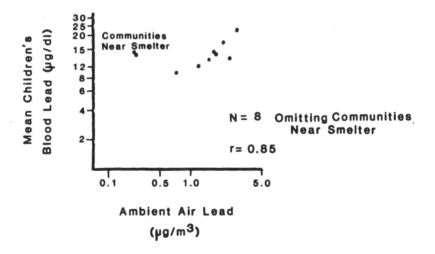

FIGURE 1. Ambient air lead levels and blood levels in third grade children in California, by sex for two northern California communities near a recently shut-down smelter, and for southern California communities with varying exposure to motor vehicular pollution.

modeling, the source of the toxic (for horses) exposures could not have been identified. Indeed, we feel that such experience justifies the support of what can be called "ecotoxicity" of pollutants. This term means the experimental study of the possible damage to animal and plant communities from pollution. Ecology is the precise analogy of epidemiology with respect to nonhuman species. So, such a word as eco-epidemiology is redundant. Indeed, leading epidemiologists have described epidemiology as human ecology.

TOXIC WASTE DUMPS: A PROBLEM IN NEED OF ECOLOGICAL ANALYSIS

Chemical and metallurgical industries produce wastes, and something must be done with them. The cheapest thing to do with them is to truck them to some apparently useless piece of ground and then to bury them. That after all is what most communities do with their domestic wastes.

But, at least in by-gone days, domestic waste was mostly food waste and paper or fabric. These materials are biodegradable, meaning that soil biology is sufficient to decompose these materials to harmless or even useful chemical fragments. The actual degradation process is a combination of the effects of water and bacteria or molds, and is what we try to replicate in making a compost heap for the disposal of garden and household waste. As the use of synthetic materials has increased, the emergence of plastic as a substantial fraction of domestic waste has changed this, since most plastic is not biodegradable.

Modern industry now produces immense amounts of nondegradable chemicals, and inevitably a portion of what they make is waste and must be dumped. To this problem is added a portion of possibly degradable but toxic material, including cyanides from electroplating, mercury, arsenic, lead and cadmium from nonferrous smelters and metallurgical plants, acid sludges from oil refineries, sludges from various chemical processes, outdated pharmaceutical agents, and the list goes on and on.

If indeed the places or waters in which they are dumped are "useless" to man at the time the dumping occurs, they are not "useless" to other living species. Neither are they always going to be "useless" to man.

We knew that mercury was toxic to fungi and algal organisms and it was used to prevent their growth, but what we did not know a few years ago was that inorganic mercury wastes could be transformed by organisms in marine muds to organic mercury compounds. These compounds entered the marine food chain, and were accumulated by fish. When the fish were consumed by humans, a severe form of poisoning involving the reproductive and nervous system was produced. It has been called "Minimata" disease after the Japanese bay where it was first detected.[2]

As cities expand, many of the apparently useless pieces of land on which wastes were dumped became desirable building sites. The new owners of the property and then the residents of the houses built there may have had no information as to the nature and status of the material which had been dumped years before. Only after the passage of some time did the odor of chemical vapors reveal to the new owners the previous use of their new homesite as a petrochemical dump and the potential this toxic material had for harm.

Now, scattered all over the country are former dump sites on or near which homes have been built. These pose a new type of community pollution problem, a historical pollution problem. The issue is not how to stop emissions, since the dumping was in most cases in the past. The issues which face these communities are whether it is safe to continue to live in these homes, if not, who and how will the householders be compensated, and how can these sites be made usable for some purpose without health risks. Federal "superfund" appropriations reflect the immense cost of cleanup of such sites.

In other communities, dumping continues to go on, and the question is how to avoid present or future toxic risks.

This set of problems has been dealt with more fully by Spivey and Coulson,[3] and only the outlines can be given here. Of major concern are the risks of birth defects in children born to families living near or over former waste dumps. Major birth defects affect about 1% of babies. As Goldsmith, Israeli, and Elkins have shown,[4] if 100 births in a cluster of households are available for study, the rate of serious birth defects would have to be increased *over ten times in order to be found to be statistically significant by conventional criteria.* If the unfavorable outcome to be looked for were something that is more frequent, such as spontaneous abortion, which affects about 10% of pregnancies, then in 100 pregnancies, the frequencies would only have to be a bit more than 2 1/2 times as high to be considered as statistically significantly increased. The general approach has been presented by Frerichs[5] in his discussion of epidemiological monitoring to see if there were any adverse effects of waste water reuse.

The difficulty is that current approaches are unlikely to be able to find a small effect or a relatively uncommon effect in populations of the size that are likely to be available for study; most such residential clusters do not have very large populations, so the number of 100 pregnancies which we have used for this example is realistic.

If there are substantial levels of toxic materials, it is likely that other forms of life with shorter lifespans and the possibility for experimental study will also be affected. It would therefore follow that ecological studies and ecotoxicity may be of greater help than a program of epidemiological study of a relatively insensitive type of outcome alone. Certainly, the epidemiological study must be done, but using conventional criteria, the risk that they will give a falsely negative result is great. Parallel ecological studies can provide a measure of protection against that risk.

There is in fact an *imperative need for ecological studies of many environmental pollution problems in parallel with community epidemiological studies.* Damage to vegetation, to poultry raising, and to livestock were part of the program of evaluation of photochemical pollution when the nature of the problem was first discovered.[6] The discovery itself of the importance of the photochemical process was actually a result of research on plant growth hormones by Professor Arie J. Haagen-Smit at California Institute of Technology in Pasadena.

The costs of cleaning up of waste dumps have already been underwritten by Congress to the extent of authorizing many billions of dollars of what is called "superfund" appropriations. To fail to try to assure that the effort is successful on both ecological and epidemiological grounds would be irresponsible. Yet, there is a serious risk that present epidemiological approaches may fail to clearly identify the nature of human health risks. A well-planned program of ecological studies is a good way to prevent overlooking some biological process that is relevant to man's health and well being. Such programs can also provide a sound basis for monitoring the soundness of waste management programs which are essential to an increasing technologically oriented society.

The health of the human community is dependent on the health of the community of other species. Ecology and epidemiology are interdependent approaches to protecting life from technological hazards.

REFERENCES

1. **Goldsmith, J. R.,** *Atmos. Environ.,* 17, 2365, 1983.
2. **Friberg, L. and Nordberg, G. F.,** in *Mercury in the Environment,* Friberg, L. and Vostal, J., Eds., CRC Press, Boca Raton, Fla., 1972, 113.
3. **Spivey, G. H. and Coulson, A. H.,** *Environmental Epidemiology: Principles and Practice,* Ann Arbor Science Press, Ann Arbor, Mich., 1984.

4. **Goldsmith, J. R., Israeli, R., and Elkins, J.,** Monitoring of unfavorable reproduction outcome (URO) among occupationally exposed groups, *Sci. Total Environ.*, 32, 321, 1984.
5. **Frerichs, R. R.,** Epidemiological monitoring of possible health reactions of wastewater reuse, *Sci. Total Environ.*, 32, 353, 1984.
6. **Stern, A., Ed.,** *Air Pollution*, Vol. 2, 3rd ed., Academic Press, New York, 1977.

Chapter 24

HISTORICAL POSTSCRIPT

John R. Goldsmith

The late Arnold Toynbee, at the time dean of historians, in 1971 wrote* concerning the next 10 years (that is until 1981): "two things at least do seem probable — the population explosion is going to continue, especially in the developing countries, and in the same ten years the price of our technological development is going to rise so steeply that it may become manifestly prohibitive." He implied that the price was going to have to be paid by the loss of health and happiness. He went on to affirm that within those ten years there would be a revulsion against our reckless pursuit of material gain. "The revolt will begin with mothers" he said. "When mothers find that their babies have been poisoned before birth by man-made poisons that have penetrated the mothers' own bodies the mothers will revolt *en masse.*" "Is it not an atrocity ?" Toynbee asks rhetorically, "to produce children who will have been condemned in advance to lifelong suffering and misery by the pollution of the world that is mankind's habitat? Is it not also an atrocity to produce children who will be condemned to lifelong poverty by the population explosion in the economically backward majority of countries." He concludes "Let us face the truth that we do not start free from encumbrance. Every generation and every individual inherits the burden of Karma, the consequences of earlier actions. We have it in our power either to mitigate our inherited Karma or to aggravate it, but we cannot jump clear of it, and we ignore it at our own peril."

To our students, we pray that our experiences and judgment may help you mitigate your burdens as citizens, parents, and possibly as scientists.

* *San Francisco Chronicle,* January 17, 1971.

Index

INDEX

F

G

O

P

Printed in the United States
by Baker & Taylor Publisher Services